ECONOMIC & FINANCIAL ANALYSIS for ENGINEERING & PROJECT MANAGEMENT

ECONOMIC & FINANCIAL ANALYSIS for ENGINEERING & PROJECT MANAGEMENT

Abol Ardalan, D.Sc.

Graduate School of Management & Technology
University of Maryland University College

CRC Press
Taylor & Francis Group
Boca Raton London New York

CRC Press is an imprint of the
Taylor & Francis Group, an **informa** business

CRC Press
Taylor & Francis Group
6000 Broken Sound Parkway NW, Suite 300
Boca Raton, FL 33487-2742

First issued in paperback 2019

ISBN-13: 978-0-367-39938-2

Economic and Financial Analysis for Engineering and Project Management

Main entry under title:
    Economic and Financial Analysis for Engineering and Project Management

A Technomic Publishing Company book
Bibliography: p.
Includes index p. 219

Library of Congress Catalog Card No. 99-65999

Visit the Taylor & Francis Web site at
http://www.taylorandfrancis.com

and the CRC Press Web site at
http://www.crcpress.com

Table of Contents

Preface

This book is about "Engineering Economics" which is a subset of microeconomics. It covers the economics and financial analyses needed to assess the viability of projects and to choose the best among many. As such it is relevant not only to engineers but also to anyone who needs to make an investment decision. For the past six years, its contents have been used as teaching material for over 400 hundred students in many disciplines ranging from electrical and mechanical engineering to information system and environmental sciences. The present version has been corrected using the students' and instructor's input. The following are some features of this book:

*. It is written clearly in an easily understandable format. It can be used in engineering and financial curricula and for self-use by financial and engineering managers.

*. It emphasizes computer application. Personal computers are now a household appliance and are used by almost all professionals. But, there are still some managers who are more familiar with the interest rate tables. To benefit those who are not familiar with computer applications, examples are also illustrated in the traditional way by using compound interest rate tables, followed by the application of computer spreadsheets. Although Quattro Pro and Excel applications are used in this book, any of the available spreadsheets can be used in a similar manner to solve the problems.

*.   It treats the section on replacement from a new perspective that is more applicable and

easier to understand than the approaches used in other books in application to real-life

cases. The spreadsheet application of this approach facilitates the sensitivity analysis

of the outcome to the input parameters in the replacement analysis.

*.   The chapter on lifetime includes the application of the lifetime system benefits and is

called lifetime worth. System benefits are an important factor in decision-making

processes involving alternatives and in the treatment of replacement decisions. It is

therefore appropriate and timely to use lifetime worth in place of life cycle cost. The

section on estimating includes estimating the cost of developing software systems.

How this book is organized

Part 1: Financial Analysis and Choice of Alternatives
Chapters 1 to 9

Financial analysis methodologies and their applications in evaluating projects and analyzing

system selection problems are presented in the first eight chapters. The ninth chapter presents

a generic approach to problem solving and a set of problems that can be solved by any or all

of the methods of financial analysis the instructor chooses. The applicability of different

methods to different situations, inflation, and tax effects is also discussed. This part of the

book establishes the basis for analysis of the subjects discussed in the subsequent parts.

Part 2: Lifetime Worth (LTW) Estimation and Calculation

Chapters 10 to 12

Lifetime analysis, which is the mainstay of the financial evaluation of alternatives and replacement decision making, is presented here. The different methodologies used to estimate the lifetime worth (LTW) and the work breakdown structure as the foundation for the calculations are discussed in this part. Special note is made of the changes in factors affecting the productivity and efficiency of the system with age. The fact that costs and benefits occur at different points in time and influence the calculation of the LTW (i.e., time value of money) is taken into account. This fact is often forgotten in most treatments of this subject, especially in cases involving estimation. The determination of the product unit cost including fixed cost and variable cost is treated in this section. A section of this part is dedicated to the discussion of the lifetime analysis of software systems, focusing on their development cost.

Part 3: Economic Life--Retirement and Replacement

Chapters 13 and 14

The different points of view in defining life of a system, system economic life, and its calculation are discussed in this part. The needs for system retirement or replacement, methodologies of replacement decision making, and the dependence of this decision on the planning horizon are discussed. Hand calculation of economic life and replacement, especially when tax and depreciation are involved, is a long, cumbersome effort. The use of computer spreadsheets makes this calculation very easy. The application of spreadsheets for this purpose is illustrated in this part.

Throughout the book, solutions to the examples are calculated using the traditional method of using compound interest rate tables and computer spreadsheets. As we will see, using spreadsheets makes calculating results not only faster but also a lot easier. Using spreadsheets also eliminates the need for some of the formulae and expressions. For example, calculations for the net present worth of uniform series, arithmetic, and geometric gradients can be performed by using the same spreadsheets constructed for determining the net present worth.

This book is designed for one semester of graduate study. Managers making system selection involving multiple alternatives and/or replacement decisions will find it useful. Engineers involved in comparative design analysis, logisticians, environmentalists, controllers, marketers, and everyone involved in decision making based on financial analysis will also find the contents of this book to be valuable in fine-tuning their decision-making process.

Acknowledgement

First and foremost, my thanks go to my many students who went through the manuscript, made comments, and at the same time encouraged me to publish the book. Without their encouragement, I would not have had the incentive to convert my class notes into book form. Dr. John Aje, director of technology and engineering programs of the University of Maryland University College, went through the original version and made useful suggestions. I appreciate his help and his support during my teaching assignments at the University College. My thanks also go to my professor and mentor, Professor John Rassmusen of The George Washington University, who has given me encouragement in everything that I start. My wife Mahvash and my children, Faroukh, Babak, and Roshan, deserve to be thanked for their technical and moral support.

PART 1

FINANCIAL ANALYSIS AND CHOICE
OF ALTERNATIVES

CHAPTER 1

Introduction

Economists, engineering managers, project managers, and indeed any person involved in decision making must be able to analyze the financial outcome of his or her decision. The decision is based on analyzing and evaluating the activities involved in producing the outcome of the project. These activities have either a cost or a benefit. Financial analysis gives us the tools to perform this evaluation. Often the decision to make is to proceed or not to proceed with a project. In cases involving investment, we want to know if the project is economically viable in order to proceed. In effect, we compare the net benefit of proceeding with the project against the consequences (good or bad) of not proceeding (the null alternative). Sometimes we are confronted with two or more courses of action (alternatives); in this situation, we want to know which alternative produces the greatest net benefit.

To be able to make the go/no go decision or to compare different projects, systems, or courses of action, we have to find a common measure to reflect all the costs and benefits and their time of occurrence. Financial analysis methods will give us this capability.

Another use of financial analysis is for the purpose of securing financing or credit for implementation of a project. Investors want to know what benefits, if any, they can gain from investing in the project. Creditors need to know if the project is secure enough so that they can get their money back. The project manager has to be able to provide them with a credible financial analysis of the project.

The first step is to identify and construct the time profile of the incomes and expenditures, namely, the cash flow diagram. The financial analysis tools will then provide a common measure to test the economic viability of the project or to compare the outcomes of the alternative decisions.

COSTS AND BENEFITS

Any project whether it is relatively simple, such as purchasing and operating a taxicab, or complex, such as creating and operating a sophisticated global communication network, involves costs and benefits. At the beginning of a project an investment is usually required. Equipment has to be purchased, buildings have to be constructed, and a host of other activities have to be conducted. All of these activities require expenditure of resources. As soon as the project gets into its operating phase, some function, such as producing a product or performing a service, that has tangible or intangible costs and benefits will be performed. If the system is producing an end item, such as a factory producing motor vehicles, then the revenue received by selling the produced items is the benefit obtained. If the project is an energy producing system, such as a utility power station, then it receives income by selling its generated power. The benefits or costs are not always given directly in monetary terms but they can be converted to monetary terms for comparison purposes. For example, in the case of a public project such as a new highway, even if there are no tolls taken, time saved, lives saved, or the convenience received by the users can be transferred into monetary benefits. In the case of defense systems, the value received is deterrence or national security which is not easily measurable in monetary terms.

Costs and benefits do not always occur at one time; they occur at different points of time during the life of the project. In most cases, the lifetime worth, that is the lifetime aggregate of all the costs and benefits, taking into account the time of their occurrence, is used to compare different projects and to decide which alternative to choose.

Let us look at the example of a housing development project. The cost items are purchase of land and design and construction of the sites and buildings. As soon as the developer starts selling the houses, benefits start coming in. Since all of the houses are not sold at the same time, the developer has to bear the cost of maintaining the unsold buildings until all of the houses are sold. This is the end of the project. The aggregate of these costs and benefits, taking into account the time of their occurrence, is the lifetime worth of this project that will determine its economic viability.

IMPORTANCE OF TIME (TIME VALUE OF MONEY)

The costs are paid and the benefits are received during different periods of the life of the system. Money can have different values at different times. This is because money can be used to earn more money between the different instances of time. Obviously, $10,000 now is worth more than $10,000 a year from now even if there is no inflation. This is because it can earn money during the interval. One could deposit the money in the bank and earn interest on it. This is the earning power of money over time and is called time value of money, that is, $10,000 now has more value than $10,000 six months from now.

Because the interest rate is the more identifiable and accepted measure of the earning power of money, it is usually accepted as the time value of money and indication of its earning power. We have to be careful not to confuse the earning power of money, which is related to interest rate, with the <u>buying power</u> of money, which is related to inflation. Inflation will be discussed later.

INTEREST RATE

When money is borrowed, it has to be paid back. In addition to the amount of the loan, an extra amount of money is paid to the lender for the use of money during the period of a loan, just as you pay a rent on a house or a car. The rate of interest i is the percentage of the money you pay for its use over a time period. The interest rate is referred to by different names such as rent, cost of money, and value of money. In investment terminology, it is called the minimum acceptable rate of return or MARR (Chapter 5). If you borrow A dollars at yearly interest rate i, at the end of the year, the interest is Ai, and the total amount you have to pay back to the lender is $A+Ai$.

To compare the value of money at different points in time, we need to use an acceptable interest rate. The interest rate will depend on the position in time that the money is needed and the length of time it is required. If money is borrowed for a long period, then the uncertainty of the economy will introduce a risk factor and influence the interest rate. For short periods, it can be assumed that the economy is stable and the risk is predictable.

Availability of money in the financial market also has an effect on the interest rate. If the banks have more money than people need to borrow, then the interest rate is low and vice versa. Money, like any other commodity, obeys the laws of supply and demand.

Interest rate is closely related to the period to which it applies. An interest rate of 1% per month is <u>NOT</u> equivalent to 12% per year.

CONTINUOUS COMPOUNDING OF INTEREST

If the interest rate for a period, e.g., one year, is i, then a loan of A dollars should obtain an interest of Ai at the end of the period. The amount of money given back at the end of the period is

$$Ai + A = A(1+i) \tag{1.1}$$

If the borrower keeps the money for another year, then he has to pay interest on $A(1+i)$ dollars. Hence, at the end of the second year he owes

$$[A(1+i) \times (1+i)] = A(1+i)^2 \tag{1.2}$$

If we continue doing this, we obtain the general equation of compounded interest:

$$A_n = A(1+i)^n \tag{1.3}$$

Where A_n is the total of the loan plus interest gained to the end of period n.

EFFECTIVE RATE OF INTEREST

From equation 1.3, we can see that the interest gained over n periods is

$$A_n\text{-A} \quad \text{or} \quad A(1+i)^n - A$$

Since the original investment was A, the interest rate is

$$i_n = \frac{A(1+i)^n - A}{A} = [(1+i)^n - 1] \tag{1.4}$$

This is called the effective interest rate, and equation 1.4 is written as

$$i_{eff} = (1+i)^n - 1 \qquad (1.5)$$

Example 1.1

A local bank announces that a deposit over $1,000 will receive a monthly interest of 0.5%. If you leave $10,000 in this account, how much would you have at the end of one year?

According to equation 1.3,

$$A_{12} = 10,000 \ (1+0.005)^{12} = 10,000 \ (1.062) = 10,620$$

This means that over a one-year period, $620 has been added to our money, which is the same as a 6.2% annual interest rate. We can see that this is not 12 times the monthly interest rate of 0.5% which is 6%. The difference between the 6.2% and 6% rates is the result of compounding monthly rather than annually.

NOMINAL INTEREST RATE

The nominal interest rate is the annual interest rate divided by the number of compounding periods in the year. If the yearly rate of interest is compounded quarterly, the quarterly nominal interest rate is the yearly interest rate divided by 4.

$$i_{Nominal} = \frac{i_{Annual}}{Number, of, Periods} \qquad (1.6)$$

Example 1.2

The annual interest rate is 6%, and the interest is compounded quarterly. What is the quarterly nominal interest rate? What is the effective annual interest rate if compounded quarterly?

Nominal interest rate $\qquad$ $I_{nominal} = 6/4 = 1.5\%$

Effective interest rate $\qquad$ $I_{eff} = (1+0.015)^4 - 1 = 6.13\%$

If the interest was compounded monthly instead of quarterly, the nominal rate would be $6/12$ = 0.5%. As we saw in example 1.1, this would yield an effective interest rate of 6.2% which is higher than what was obtained by compounding quarterly.

EQUIVALENCE

Making a decision on multiple alternatives requires a common measure of performance. Costs and benefits occur at different points in time and, hence, have different values. Financial analysis methods are tools that will enable us to evaluate the aggregate of these costs and benefits with a common measure. We will see later that these common measures are

Net present worth

Net future worth

Benefit - cost ratio

Equivalent Uniform Annual Worth

Rate of return

Two or more projects are economically equivalent if they have the same result when measured by any of the above measures, that is, the economical consequence of either project is the same. A common interest rate should be used in the process of measurement. Projects that are equivalent at one interest rate are not necessarily equivalent when another interest rate is used in the measurement.

CASH FLOW DIAGRAM

The graphic presentation of the costs and benefits over the time is called the cash flow diagram. This is the time profile of all the costs and benefits. It is a presentation of what costs have to be incurred and what benefits are received at all points in time.

The following conventions are used in the construction of the cash flow diagram:

* The horizontal axis represents time

* The vertical axis represents costs and benefits

* Costs are shown by downward arrows

* Benefits are shown by upward arrows

● All the benefits and/or costs incurred during a period are assumed to have been incurred at the end of that period. Since the period is normally a year, this is called the "end of the year" rule.

Example 1.3

A car leasing company buys a car from a wholesaler for $24,000 and leases it to a customer for four years at $5,000 per year. Since the maintenance is not included in the lease, the leasing company has to spend $400 per year in servicing the car. At the end of the four years, the leasing company takes back the car and sells it to a secondhand car dealer for

$15,000. For the moment, in constructing the cash flow diagram, we will not consider tax, inflation, and depreciation.

Step 1:

Draw the horizontal axis to represent 1,2,3, and 4 years.

Step 2:

At time zero, i.e., the beginning of year 1, the leasing company spends $24,000. Hence, at time zero, on the horizontal axis, a downward arrow represents this number.

Step 3:

At the end of year 1, the company receives $5,000 from his customer. This is represented by an upward arrow at the end of year 1. The customer also spends $400 for maintaining the car; this is represented by a downward arrow.

The situations at years 2 and 3 are exactly the same as year 1 and are the presentations on the cash flow diagram exactly as for the first year.

Step 4:

At the end of the fourth year, in addition to the income and the expenditure as in the previous years, the leasing company receives $15,000 by selling the car. This additional income is represented by an upward arrow.

The project ends at this time, so we have nothing else to insert in the cash flow diagram. We have represented all the costs and benefits in the cash flow. At this point, it is a good idea to go back through the life of the project and make sure that nothing as expressed in the description of the project is left out.

Fig. 1.1 represents the cash flow diagram of this project and is the financial model of this project.

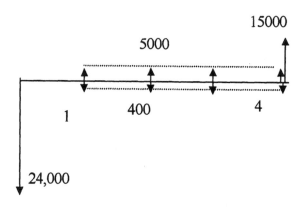

Fig. 1.1

The costs and benefits for each year can be deducted from each other to present a "netted" cash flow diagram as presented in Fig. 1.2

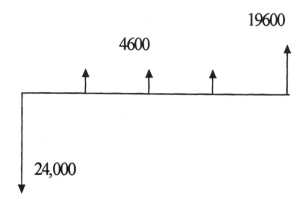

Fig. 1.2

INCLUSION OF NONMONETARY COSTS AND BENEFITS

In the above example, all costs and benefits were indicated by dollar amounts. Indeed, to be able to perform financial analysis, all costs and benefits should be entered into the analysis in monetary values. But, we know in practice that some benefits are not easily convertible to monetary values. For example, a road going through scenic areas has nonmonetary values to the people driving through it. How do we account for this in our analysis? How do we account for the beautiful façade of a downtown office building in a major city? The so-called WOW factor. In some cases, a good subjective value can be attached to these factors. For example, in the case of the building, we can estimate how many more tenants would volunteer to have an office with a good façade, and how much more rent they would be willing to pay. The additional cost of making a highway safe can be estimated by the value of the lives that can be saved. A reasonably good assessment can be made using the following expression: *the value of anything can be estimated by measuring the cost of not having it.* If you can take an offer of $x not to go and see a ball game on Saturday, it means that to you the value of going to see that game or the cost of not going to the game is x dollars.

IMPORTANCE OF CASH FLOW DIAGRAM

The cash flow diagram is the most important and essential element of financial analysis. A proper and accurate cash flow diagram should be constructed and tested before an attempt is made to perform the financial analysis. Indeed, with today's special handheld calculators and personal computer spreadsheets, the financial analysis is completed very quickly without much financial knowledge required on the part of the operator. But, the construction of a cash flow diagram requires a deep understanding of the financial situation of the project or problem at hand. No computer can provide the right answer if the cash flow diagram is not constructed properly and accurately.

All the cost and benefit components occurring during the course of the project and their time of occurrence should be accurately presented in the cash flow diagram. Any costs related to this project incurred before the zero time of the analysis are considered "sunk cost" and do not enter in the analysis. This is a very important point to remember. It does not mean that in our future activity we should not consider taking a course of action to recover the sunk cost. It means that the fact that we have spent money up to this point should not cause us to continue a non-profitable project; in other words, " don't send good money after bad money". The interest rate i is assumed to be constant for the duration of the project or operation of the system under analysis. A typical cash flow diagram is presented in Fig. 1.3

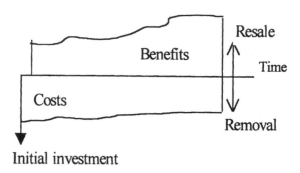

Fig. 1.3

THE PROCESS OF DECISION MAKING

As mentioned previously, the main reason for going through the financial analysis is to make a decision. The following are the points to consider and the process to go through (not necessarily in the order given) in making the decision.

* Objective (what is your objective, what do you want to achieve)

* Viewpoint (from whose point of view you are looking at the problem)

* Criteria (Objective Function, what determines the achievement, what is success)

* Alternatives (how many reasonable roads to the objective)

* Constraints (what are the limitations in resources, actions, etc.)

* Consequences of the alternatives over time (prediction of the outcomes)

* Planning Horizon (how far ahead are you thinking, for what period of time?)

* The Model (a mathematical or graphical representation of the elements of the problem and their interactions)

* Differential Consequences (the differences between the outcomes of the alternatives)

* Risk and Uncertainty (what effects do the uncontrollable elements have in the outcome)

* Opportunity Cost (what is forfeited when the actions are taken)

* Objective Function (which alternative optimizes the "Objective Function")

FINANCIAL ANALYSIS METHODS

Several analysis methods could be used to evaluate the economic viability of a project or to compare the financial merits of several projects. The same analysis techniques can be used to calculate the lifetime worth of a system and perform the replacement analysis.

All financial analysis methods have one thing in common. They attempt to find a common measure for the aggregate of the costs and benefits so that the net outcome of the project, negative or positive, is measured and that the comparison of the alternatives is based on a common measure and criteria, always considering the time value of money. Construction of the cash flow diagram is the first and essential part of financial analysis.

DERIVATION OF THE FORMULAE

Financial analysts only apply the tools and do not remember or care how the equations were derived. Therefore, emphasis in this book is on the application of the formulae rather than their derivation. Interested readers can find the derivations in the appendix to this book.

PROBLEMS

1-In a housing project the following sequence of events occurs.

- At the start of the project (time zero), land is bought at $1,000,000

- Two months later, $100,000 is paid to the architect for preparing the design

- In month 4, construction is started and the cost of construction (labor and material) is $150,000 per month

- Every month, one house is built (a total of 12 houses); the first one is ready for sale in month 6

- During every month starting from month 8, one house is sold for a price of $150,000 each

- After all of the houses are built and before all are sold, the cost of maintaining the site is $10,000 per month

Draw the cash flow diagram.

2-Mr. Shop purchases a pizza shop for $120,000. Its operation will result in a net income of $15,000/Yr for the first year, increasing by $2,000 each year after year 1. At the end of the fifth year, the shop is sold for $155,000. Draw the cash flow diagram for this project.

3- A credit card company announces that its interest rate is 1.5% per month.

What is the corresponding effective annual interest rate?

4-Your local bank has a promotional saving program that pays an interest rate of 6% per year compounded monthly. If you deposit $1,000 on January 1 in this bank, how much will you have in your account at the end of year 1 and year 2?

5- Mr. X deposited $1,500 in a savings account at the local bank and went on assignment overseas. After two years, he returned and noticed he had $1,800 in his account. What annual effective rates of interest had the bank given him if they compounded the interest quarterly? What if they compounded annually?

6- The local bank advertised an investment program with an annual 16% interest rate, compounded quarterly. You can also choose to invest your money at the local branch of an out-of-town bank that will give you an annual interest rate of 17.5%. How much more will you gain or lose per year if you invest $1,000 at the local bank instead of at the out-of-town bank?

CHAPTER 2

Present Worth

PRESENT VALUE (PV)

From the discussion on interest rate (Chapter 1), we can conclude that $1,000 lent at a 12% rate for one year will earn $120 and will add to the original $1,000 to total $1,120. Therefore, at a 12% interest rate, $1,000 is now the equivalent of $1,120 a year from now and is called the present value of $1,120. The present value is therefore a function of time t and interest rate i.

$$PV = f(t,i) \qquad (2.1)$$

In the present value method, the present time (time zero or start of year 1) equivalent value of all the costs and benefits incurred during the life of the system or the project is calculated using a specific interest rate. The $1,000 is sometimes called the discounted value of the $1,120, since $1,120 a year from now was discounted (reduced) to $1,000 at present.

Equation 2.2a gives the equivalent present value of a future value (cost or benefit) at n equal consequent intervals of time t from present time with the constant interest rate i per interval prevalent during the total time nt.

$$PV = FV \times (1+i)^{-n} \qquad (2.2a)$$

The above equation is represented by the expression

$$P = F(P/F,i,n) \qquad (2.2b)$$

$$PV = P = \text{Present Value}$$

$$FV = F = \text{Future Value}$$

Usually n is the number of years, t is one year, and i is the annual interest rate. The multiplier of the future value in expression 2.2b normally shown as (P/F, i, n), is called the <u>single payment, present value factor,</u> and its values for any i and n are given in the appropriate compound interest tables.

NET PRESENT WORTH

The net difference of the present costs and benefits is the net present worth.

$$NPW = PV \text{ (Benefits)-PV (Costs)}$$

When we calculate this value for all the benefits and costs and for all the years of the system life, the system lifetime worth is obtained. The use of the present value refers all the costs and benefits to a single point in time (present) so a just comparison between systems can be made.

PRESENT VALUE OF UNIFORM ANNUAL SERIES

If the same benefits and/or costs occur for every period, e.g., every year as in Fig. 2.1, then in this case, the present worth is called <u>uniform series present worth factor.</u>

Where A is the annual cost or benefit, the present value of the uniform series can be calculated by multiplying A and equation 2.3a or can be obtained from the compound interest tables using expression 2.3b.

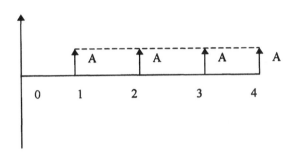

Fig. 2.1

$$\frac{(1+i)^n - 1}{i(1+i)^n} \qquad (2.3a)$$

$$PV = P = A(P/A, i, n) \qquad (2.3\ b)$$

Example 2.1

Calculate the net present value of the leasing project of example 1.3 using the netted cash flow diagram of Fig. 2.2a and assuming an interest rate of 10%.

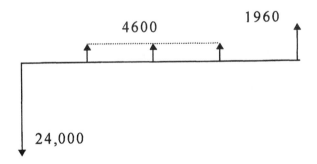

Fig. 2.2a

We have to calculate the preset (time zero) value of all the costs and benefits. The simplest and therefore the longest way to do this is to break the above cash flow into three components shown in Fig. 2.2b.

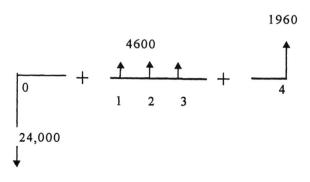

Fig. 2.2b

Step 1:

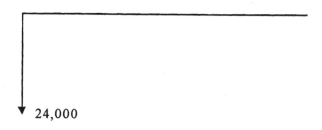

Fig. 2.2c

The $24,000 cost occurs at time zero, so its present value is $24,000. Now, we have to obtain the NPW of the other elements of the cash flow.

Step 2:

      The $4600 benefit is received for three years (n=3). We can then use equation 2.3a to calculate the equivalent value of the total of these three payments referred to at time zero.

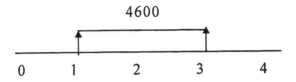

4600

0   1   2   3   4

Fig. 2.2d

i is 10%, that is 0.1 and n=3, hence present value at time zero is

$$PV = \frac{(1+0.1)^3-1}{0.1(1+0.1)^3}=4600$$

$$PV = \frac{0.331}{0.1331} \ 4600 = 11439.52$$

We could have used expression 2.3b.

$$A = 4600$$

From the compound interest tables, the page for i =10% , row n=3 and column for P/A, we obtain

$$(P/A, 1, n) = 2.487$$

Therefore, $PV = 4600 * 2.487 = 11440$

This is the same value calculated before. The small difference is due to rounding of the numbers.

Step 3:

Calculate the present value of the $19,600 at year 4. For this, we use equation 2.2a.

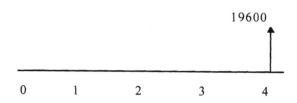

Fig. 2.2e

$PVB = 19600 * (1+0.1)^{-4} = 13387.1$

Step 4:

Calculate total cost, total benefit, and net worth.

Total Cost = 24,000

Total Benefit = 11439.52 + 13387.1 = 24826.62

Net Worth of the Project = 826.62

Again, we could have used the expression 2.3b and used the page for 10% from the compound interest rate tables. We would have obtained the same answer.

There are many different ways to arrive at this number. The above is the simplest way and is prone to less error. As you, the reader, gain more experience, you will develop shortcuts with which you are more comfortable. You will then be able to solve some problems with only one line of arithmetic. The trick is to break the netted cash flow into components with which you feel comfortable.

The Excel spreadsheet calculation of the net present worth for this problem is shown below. The Quattro Pro spreadsheet expression to be used is also shown.

	A	B	C	D	E	F	G	H	I	J	K	L	M	N
1	Example 2.1													
2			Set up the spread sheet as shown below:											
3														
4		Interest 10%												
5														
6			Year	0	1	2	3	4						
7				-24000	4600	4600	4600	19600						
8	Then:													
9		For Excel: Use =D7+NPV(C4,E7..H7) at G9					NPW=826.58							
10														
11		For Quattro Pro: Use (@NPV(C4,D7..H7) at H11 NPW=826.58												
12														

PRESENT VALUE OF ARITHMETIC GRADIENT SERIES

These are annual series with constant increasing values such that

$$A_1 = 0$$

$$A_2 = A_1 + G = G$$

$$A_3 = A_2 + G = 2G$$

$$A_n = A_{(n-1)} + G = (n-1) G$$

The cash flow diagram is shown in Fig. 2.2, and G is called the gradient.

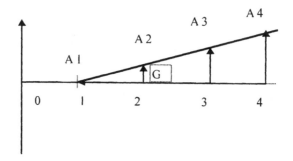

Fig. 2.3

The present value is

$$P = G\frac{(1+i)^n - in - 1}{i^2(1+i)^n} \qquad (2.4a)$$

$$P = G(P/G,i,n) \qquad (2.4b)$$

As in the other expressions, the value of (P/G, i, n) can be found in the compound interest tables.

Example 2.2

A project has a net income of $50 the first year, increasing by $100 every year for the next three years. What is the net present worth of this project at an interest rate of 10%?

In this example, the annual income is determined according to an arithmetic series, and we have to apply equation 2.4a or expression 2.4b.

The following cash flow diagram presents the model for this example.

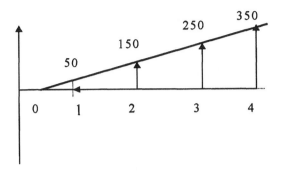

Fig. 2.4a

The above cash flow can be broken into two cash flows as in Fig. 2.4b and the calculation is

PV = G (P/G, i, n) + A (P/A, i, n) Where G = 100, A = 50, and n = 4

PV = 100 (P/G, 10%, 4) + 50 (P/A, 10%, 4) = 100 (4.378) + 50 (3.170)

= <u>596.3</u>

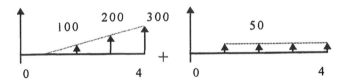

Fig. 2.4b

The spreadsheet calculation is shown below. As can be seen we used the same spreadsheet configuration in this case as for Example 1.2. In using the compound interest table, we had to break the cash flow diagram into two components and use the P/G and P/A expressions. The use of the spreadsheet made the operation a lot simpler.

	A	B	C	D	E	F	G	H	I	J	K	L	M	N
1	Example 2.2													
2			Set up the spread sheet as shown below:											
3														
4		Interest 10%												
5														
6			Year	0	1	2	3	4						
7				0	50	150	250	350						
8	Then:													
9		For Excel: Use =D7+NPV(C4,E7..H7) at G9						NPW=596.30						
10														
11		For Quattro Pro: Use (@NPV(C4,D7..H7) at H11 NPW=596.30												
12														

MULTIPLE ALTERNATIVES AND EQUALIZING LIVES

In comparing several systems or projects with equal lives, we calculate the NPW for all and choose the one with the highest NPW. If the lives of two (i.e., n_1 and n_2) are not equal, the project with the longer life is still going on while the project with the shorter life is terminated, an inconsistency in the analysis is presented. What do we do with the resulting money (NPW) from one project while the other one is ongoing? To address this inconsistency, we will assume that the project with the shorter life will repeat n_1 / n_2 times. In this process, we equalize the length of the cash flow diagram for both alternatives. If n_1 and n_2 are not divisible, we repeat both of them as many times as needed to make the analysis life equal to n, where the common analysis life "n" is the <u>least common multiplier</u> of the two lives

n_1 and n_2. This is called "equalizing the lives". We construct the net cash flow

diagrams of the two systems for life n and calculate the net present values of the

resulting cash flows. In this manner, we ensure that the effect of the extra life of

one system over the other and time value of money are taken into account.

Example 2.3 illustrates this methodology.

EXCLUSIVITY

In analyzing the choice between two or more alternatives, we assume that

these alternatives are <u>mutually exclusive</u>; that is, choosing one alternative

precludes the choice of all the others. We make this assumption throughout this

book.

Example 2.3

Suppose the leasing company of Example 2.1 has to choose between the following

two projects:

1. Lease the car exactly as the case in Example 2.1 which had a net present

    worth of $826.62

2. Buy a car at $40,000, lease it for two years at $12,000 per year with no

    maintenance cost, and sell it for $24,000 at the end of two years.

Assuming an interest rate of 10%, which project should we choose? In this

problem, $n_1=4$ and $n_2=2$, therefore, the least common multiplier of n_1 and n_2 is

equal to four. That means project 1 is going on for two years after project 2 ends.

Can we make a fair comparison? What does the leasing company do with the

money received from project 2? To solve this problem, we assume that they

engage in project 2 for another two years. This is the same as assuming that we

buy the same car and lease it under the same condition for another two years. The

procedure is as follows.

Step 1:

Draw the cash flow diagram of the project with the smaller initial

investment (in this case project 1) and check it against the null alternative. If NPW

was not positive, we would disregard this project and would check project 2 to see

if its NPW is greater than zero. In this case, we have already done the analysis and

know that the NPW of project 1 is positive.

Step 2:

Since $n_1=4$ and $n_2=2$, then their common multiplier is 4 so we draw the

cash flow diagram of both projects for n=4 years.

We repeat the netted cash flow diagram of project 2 as in Fig. 2.5b to get a four-

year life for this project and, hence, equalize both lives.

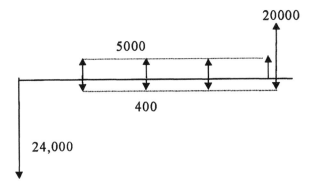

Fig. 2.5a

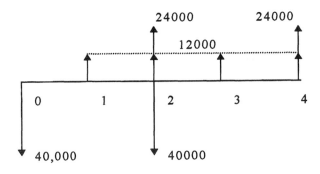

Fig. 2.5b

<u>Step 3</u>:

Draw the netted cash flow diagram of project 1 and that of extended life project 2, as show in Figs. 2.5c and 2.5d.

Now calculate the respective NPW of the two projects.

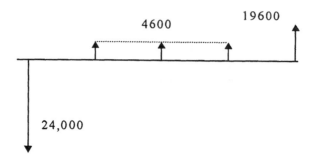

Fig. 2.5c

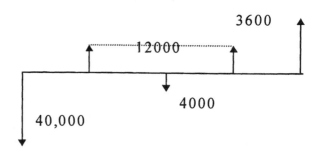

Fig. 2.5d

Step 4:

NPW 1 was calculated before and is equal to $826.58.

NPW 2 is calculated using expressions 2.2b:

= -40000 + 12000 (P/F, 10,1) - 4000 (P/F,10,2) + 12000 (P/F,10,3) + 36000

(P/F,10,4)

= -40000 + 12000 (0.9091) -4000 (0.8264) + 12000 (0.7513) + 36000 (0.6830)

= 1,207.2

Since project 2 has a higher NPW, it would be the project to choose.

The spreadsheet calculation is shown below in Example 2.3:

	A	B	C	D	E	F	G	H	I	J	K
	Example 2.3										
			Set up the spreadsheet as shown below:								
		Interest	10%								
		A: Cash Flow for project 1									
			Year	0	1	2	3	4			
			Inflow	0	5000	5000	5000	20000			
			Outflo	24000	400	400	400	400			
			Net	-24000	4600	4600	4600	19600			
	Then:										
		For Excel: Use = D10+NPV(C4,E10..H10) at J9							NPW=	826.58	
		For Quattro Pro: Use (@NPV(C4,D10..H10) at J15							NPW=	826.58	
		B: Cash Flow for project 2 with extended life to 4 years									
			Year	0	1	2	3	4			
			Inflow	0	12000	36000	12000	36000			
			Outflo	40000	0	40000	0	0			
			Net	-40000	12000	-4000	12000	36000			
	Then:										
		For Excel: Use = D21+NPV(C4,E21..H21) at J26							NPW=	1207.6	
		For Quattro Pro: Use (@NPV(C4,D21..H21) at J26							NPW=	1207.6	

PROBLEMS

1- The cash flow of an investment is shown below. What is the NPW (i=15%)?

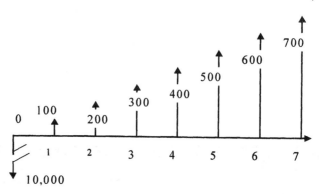

2- Mr. "X", a friend of yours, is asked to invest in the following project:

Installation and operation of a facility with a life span of five years. The initial

investment is $90M. It will have a net profit of $25M/Yr the first two years and

$30M/Yr in years 3,4, and 5. At the end of year 5, it has to be disposed of at a

cost of $10M with no resale value. If he has the money and his opportunity cost

of money is 10% (i=10%), would you advise him to invest or not? Yes? No?

Why? Explain.

3- Hosbol Corporation has purchased a system for $1 million. The net income

from operating this system is $300,000 per year. Assuming a life of five years and

no salvage value, what is the Net Present Worth (NPW) of this system (i=10%)?

4- Equipment is bought for an initial cost of $20,000. Its operation will result in a

net income of $6,000/Yr for the first year, increasing by $1,000 each year after year 1. At the end of the fifth year, the equipment is sold for $5,000. The prevailing interest rate for the next five years is estimated at 10%.

a.   Draw the cash flow diagram for this project.

b.   Calculate the NPW.

5- Production equipment is bought at an initial price of $10,000. The annual operation and maintenance cost is $100. The salvage value at the end of the 15-year life is $500. Using MARR of 10 %, calculate the net present worth. Another model of the equipment with the same initial price and annual cost brings in an income of $1,100 per year but has no salvage value at the end of its 15-year life. As an investor, would you invest in a or b? Why?

6- Board members at Darbol Corporation received two proposals for a machine they may want to purchase. They also can choose to invest their capital and receive an interest rate of 15% annually. Using the following data about the machine, what is their most economical course of action? Use the net present worth method.

Data	Machine A	Machine B
Initial Cost	$180,000	$240,000
Salvage Value	$40,000	$45,000
Annual Benefit	$75,000	$89,000
Annual Cost	$21,000	$21,000
Life	5 years	10 years

7- Members of the board at ACE Corporation received three proposals for a machine they may want to purchase. They also can choose to invest their capital and receive an interest rate of 15% annually. Using the following data about the machines, what is their most economical course of action? Use a 10-year life span.

Data	Machine A	Machine B	Machine C
Initial Cost	$180,000	$235,000	$200,000
Salvage Value	$38,300	$44,800	$14,400
Annual Benefit	$75,300	$89,000	$68,000
Annual Cost	$21,000	$21,000	$12,000

8- Mr. "X", a friend of yours, is asked to invest in either of the following two mutually exclusive projects. His MARR is 10%.

a.   A car repair system is offered with an initial cost of $30,000 and a net annual income of $15,000. The system will have a salvage value of $9,000 at the end of its three-year life.

b.   A car cleaning operation is offered with $40,000 initial cost, net annual income of $20,000 for the first three years, and $5,000 for the last three years of its life. Its salvage value at the end of its six-year life is $6,500. What do you recommend he should do?

9- A venture group is contemplating investment in either of the following projects:

a. Establish a cosmetic store with an initial cost of $100,000 and an annual net income of $20,000; the business is estimated to have a resale value of $300,000 after a four-year life.

b. Take over a beauty parlor with an $80,000 initial payment and an annual net income of $25,000 for four years. The lease will end at the end of the four years with no obligation on either side.

He will pay you $2,000 to make him a recommendation based on sound economic analysis. What would you recommend? (Assume an interest rate of 8%.)

10- A local internet provider advertises its no-time-limit service with a one-year subscription of $20 per month, two years at $11/month, and three years at $9/month. If you need to have a service from this company and your cost of money is 6%, which option do you take?

CHAPTER 3

Future Worth

FUTURE VALUE (FV)

A corollary to the present value and net present worth is the future value and the net future worth (NFW). Equation 3.1a of the future value is obtained by transposing equation 2.2a.

$$FV = PV(1+i)^n \qquad (3.1a)$$

As we saw in Chapter 1, this is also the equation for compound interest. The expression for the future value is

$$P = P \ (F/P, \ i \ n \ ) \qquad (3.1b)$$

The value of the multiplier $(F/P, \ i \ n)$ is found in the row n and column P/F of the interest rate page corresponding to interest rate i.

The net future worth is   Net Present Benefits – Net Present Costs

Example 3.1

If $1,000 is kept in a savings account that earns 6% interest, what would be the value of money in four years?

We can use either equation 3.1a, or we can use expression 3.1b and the interest rate tables.

We choose the latter.

$$FV = 1000 \, (1.262) = 1262$$

The spreadsheet solution to this example is shown below:

	A	B	C	D	E	F	G	H	I	J	K	L	M	N
1	Example 3.1													
2			Set up the spreadsheet as shown below:											
3														
4		Interest	6%											
5														
6			Year	0	1	2	3	4						
7				1000	0	0	0	0						
8	Then													
9		For Excel: Use =FV(C4,4,0,1000,1) at J9					NFW=1,262.48							
10														
11		For Quattro Pro Use @FUTV(C4,D7 F7,1) at J11 NFW=1,262 48												
12														
13														

FUTURE VALUE OF UNIFORM ANNUAL SERIES

The Equation 3.2a for the future value of a uniform annual series of A is the transposition of equation 2.3a.

$$F = A \frac{(1+i)^n - 1}{i} \qquad (3.2a)$$

The cash flow diagram corresponding to the above equation is

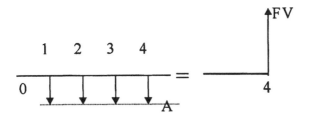

Fig. 3.1

Note that Fig. 3.1 is the reverse of Fig. 2.1 in that it brings the annual values forward.

The corresponding expression is

$$FV = A \, (F/A, \, i, \, n) \qquad (3.2 \, b)$$

The multiplier $(F/A, i, n)$ is called the sinking fund factor, i.e., $A is sunk every year with interest rate i to receive a future value (FV) n years later.

Please note that for the formula 3.2a and the expression 3.2b to be applicable we must have a value for cost or benefit (even if this value is zero) at the last year, that is year n.

Example 3.2

If we save $500 every year in a bank account that gives 6% interest, how much do we have at the end of the fifth year?

$$FV = 500 \, (F/A, \, 0.06, \, 5)$$

$$= 500 \, (5.637) = 2818.5$$

The spreadsheet calculation of the net future worth is shown below. As in the case of the present worth, both expressions to be used from Quattro Pro and Excel are shown. The small difference between the numbers is due to different rounding of the result depending on the

number of decimal point.

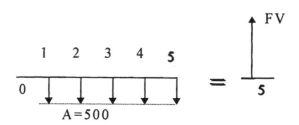

A	B	C	D	E	F	G	H	I	J	K
1	**Example 3.2**									
2		Set up the spreadsheet as shown below:								
3										
4	Interest	6%								
5										
6		Year	0	1	2	3	4	5		
7			0	500	500	500	500	500		
8	Then:									
9		For Excel: Use =FV(C4,4,0,1000,0) at J9						NFW=2,818.55		
10										
11		For Quattro Pro: Use @FUTV(C4,D7..F7) at J11					NFW=2,818.50			
12										

MULTIPLE ALTERNATIVES

The procedure for multiple alternatives is exactly as in the case of the NPW. We have to

equalize lives first and then calculate the NFW of each. The alternative of choice is the one

with the highest NFW.

PROBLEMS

1-The cash flow of an investment is shown below. What is the NFW at 15%?

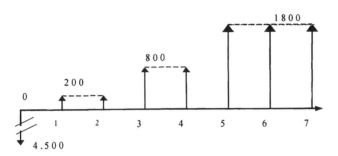

2- A 14-year-old high school student wants to save money to buy a secondhand car for $6,000 when he obtains his driver's permit at age 16. He wants to know how much per month he should save from the allowance he receives from his rich father so that he will have enough money to buy the car. (He starts saving from the first month after his birthday.) His savings account at the local bank gives an interest rate of 6% per year compounded monthly. Please help him and solve the problem for him. If the bank uses the same annual rate but compounds monthly, should he save more or less every month? Why? Explain your reasoning.

3- Mr. Futurolog, a friend of yours, has invested $9M in a fast-food franchise chain. He has had a net profit of $2.5M/Yr the first two years and $3M/Yr in years 3,4, and 5. If his cost of money is 10%, what is the minimum price he should sell the franchise for?

4- The cost of each year of college is $20,000. How much per year at an interest rate of 7% should you save for your newborn baby so that he can go to a four-year college at the age of 17?

CHAPTER 4

Annual Worth

ANNUAL WORTH AND EQUIVALENT UNIFORM ANNUAL WORTH

The annual worth is the net of all the benefits and costs incurred over a one-year period. Therefore, we present the net of all the different benefits and costs incurred at different points of time in a one-year period with one number, and we call it the annual worth. For a system whose life is longer than one year, this number will be different for different years. For systems having more than one year of life, we can calculate a single virtual number that represents an equivalent annual net benefit or cost for the duration of the system life. This virtual number is called the equivalent uniform annual worth (EUAW) and is equal to the total benefit and cost of the system as if it was spread evenly throughout the years of its life. We can express this in a different way. The net present worth of the system, calculated as if its net benefit or cost for each year was the calculated EUAW, is the same as the net present worth of the same system using the real values of costs and benefits at their real time of occurrence.

The simplest process for calculating this number takes two steps.

Step 1:

All the costs and benefits are transferred to the present year using equation 2.2a to calculate the NPW.

Step 2:

Multiplying the NPW by a factor called <u>capital recovery factor</u> converts it to EUAW. This multiplier is

$$\text{EUAW} = \text{NPW} \times \frac{i(1+i)^n}{(1+i)^n - 1} \qquad (4.1a)$$

Or as in previous cases

$$\text{EUAW} = A = P(A/P, i, n) \qquad (4.1b)$$

(A/P, i, n) is the capital recovery factor, and its value for any i and n can be found in compound interest tables. This is the same as spreading the NPW of a project over the life of the project. When the EUAW of a system or project is a positive number, it indicates that the project is economically viable or profitable. The advantage of this method is that we need not worry about the unequal lives or the unequal initial investment of the two systems being compared. They are taken into account automatically through the mathematics of the analysis method.

If the EUAW method is used for choosing among more than two alternatives, we simply have to calculate the EUAW of all of them and choose the one with the highest EUAW.

When a bank gives you a loan to buy a house, it spreads the loan over the next 15 or 30 years. In effect, your mortgage payments are the EUAW of the loan.

Example 4.1

Find the EUAW for the project of example 1.3.

Step 1:

Draw the cash flow diagram and calculate the NPW:

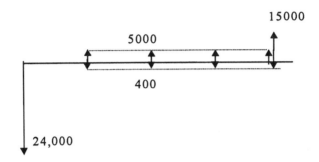

We have already done this and we know that

$$NPW = 826.61$$

We can either use equation 4.1a or expression 4.1b. Using the latter,

$$EUAW = 826.61(A/P,10,4) = 826.61 (.3155) = \underline{260.8}$$

This means that the NPW of 826.61 is the same as an annual value of 260.8 for four years.

The use of spreadsheet will give us the same result.

	A	B	C	D	E	F	G	H	I	J	K
1	**Example 4.1**										
2				Set up the spreadsheet as shown below:							
3											
4		Interest	10%								
5											
6			Year	0	1	2	3	4			
7				-24000	4600	4600	4600	###			
8											
9	First we calculate the NPW as we did in example 2.1										
10											
11			For Excel: Use =D7+NPV(C4,E7..H7) at G9						NPW=	826.58	
12			For Quattro Pro: Use (@NPV(C4,D7..H7) at H1						NPW=	826.58	
13											
14	Then using the value of the NPW as the presnt value::										
15											
16			For Excel: Use =PMT (C4,4,-J11,0,0) at J16						EUAW=	260.76	
17			For Quattro Pro: Use @PYT(C4,4,J12) at J17						EUAW=	260.76	
18											
19											

Example 4.2

A $120,000 house is bought by making a $20,000 down payment and obtaining a loan

from the local bank at an interest of 10% for 30 years. What is the annual payment?

We can use equation 4.1a or expression 4.1b to solve this problem. Using expression 4.1b

$$EUAW = A = P\,(A/P, 10, 30)$$

Using the compound interest rate tables

$$EUAW = 100,000\,(0.1061) = \underline{10,610}$$

The spreadsheet solution is

	A	B	C	D	E	F	G	H	I	J	K
1	Example 4.2										
2											
3		Interest	10%								
4			Value-Down Payment = Loan amount								
5		Loan Amount = 120000-20000 = 100000									
6											
7		Then using the loan amount as the presnt value									
8											
9			For Excel: Use =PMT (C3,30,F5,0,0) at K10							EUAW=	(10,607.92)
10			For Quattro Pro: Use @PYT(C3,30,F5) at K10							EUAW=	-10,607.90
11											

IMPORTANT POINT

Can we divide this number by 12 and obtain the monthly payment? The answer is <u>NO</u>.

The reason is that the interest rate and period of payment go together. The calculation of 30-

year monthly payments involves 360 periods. Our annual interest rate is 10%. We have to

deduce the monthly rate from the yearly rate, as we did in Chapter 1, and apply it with n=360.

Example 4.3

Calculate the EUAW of the following cash flow at 15% interest rate:

Years:	0	1	2	3	4	5	6
Payments:	100	200	-50	300	-100	200	100

First we calculate the NPW:

NPW = 100(P/F, 15%,1)+200(P/F,15%,2)-50(P/F,15%,3)+300(P/F,15%,4)

-100(P/F,15%,4)+200(P/F,15%,5)+100(P/F,15%,6) =518.8

EUAW = 518.8 (A/P,15%,6) = 137

	A	B	C	D	E	F	G	H	I	J
1	Example 4.3									
2										
3		Interest	15%							
4										
5		Year	0	1	2	3	4	5	6	
6		Payment	100	200	-50	300	-100	200	100	
7										
8	First we calculate the NPW as we did in example 2.1									
9										
10	For Excel: Use =C7+NPV(C4,D7:I7) at J11							NPW=	518.85	
11	For Quattro Pro: Use (@NPV(C4,D7..H7) at H11							NPW=	518.85	
12										
13	Then using the value of the NPW as the presnt value::									
14										
15	For Excel: Use =PMT (C4,6,-J11,0,0) at J16							EUAW=	137.10	
16	For Quattro Pro: Use @PAYT(C4,6,J12) at J17							EUAW=	137.1	

EUAW OF AN ARITHMETIC GRADIENT

The EUAW of a gradient is calculated by equation 4.2a or expression 4.2b, where G is

the gradient. The spreadsheet calculation is exactly as in Example 4.3.

$$EUAW = G \times \frac{i(1+i)^n - in - 1}{i(1+i)^n - i} \qquad\qquad (4.2a)$$

$$EUAW = A = G(A/G,i,n) \qquad\qquad (4.2b)$$

Example 4.4

An employee with an annual pay of $30,000 is told he is going to get an annual pay increase of $1,200 each year. The increase starts in the second year of his employment. What is the EUAW of his increase for the first five years at an assumed interest rate of 7%?

From the interest rate tables, we can obtain A/G for i=7% and n=5 to be

$$EUAW = G \ (A/G, \ i, \ n)$$

$$(A/G,i,n) = 1.865$$

Hence, $EUAW = 1200 * 1.865 = 2238$

We can also use equation 4.2a:

$$EUAW = 1200 \times \frac{(1+0.07)^5 - 0.07 * 5 - 1}{0.07(1+0.07)^5 - 0.07} = 2255$$

The difference of 0.08% is due to the rounding in the table.

We can also use the spreadsheet as in Example 4.3.

MULTIPLE PROJECTS

In case of multiple projects, the project with the highest EUAW is the best choice, and we do not have to equalize lives. This is one of the advantages of the EUAW comparison.

PROBLEMS

1-The cash flow of an investment is shown below. What is the EUAW? (i=10%)

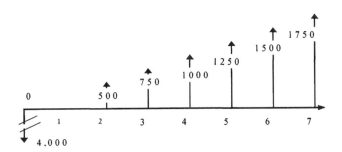

2- Greed Corporation purchased a foundry system for $1 million. The net income from operating this system is $300,000 per year. Assuming a life of five years and no salvage value, what is the EUAW of this system? Greed's cost of money is 12%.

3- Darbol Corporation received two investment proposals. The estimate of the financial situation of each proposal is presented in the following table. Darbol also has the choice of investing the capital and receiving an interest rate of 15 % annually.

Using the EUAW method, perform the financial analysis and make your recommendation as to which of the proposals, if any, they should accept.

Data	Proposal A	Proposal B
Initial Cost	$50,000	$140,000
Salvage Value	$20,000	$30,000

Annual Benefit	$30,000	$40,000
Annual Cost	$15,000	$10,000
Life	3 years	6 years

4- Mr. and Mrs. Smith, who both work for a national retail chain, purchased a house with a price of $400,000. They paid a down payment of $40,000 using their savings and took a 30-year loan from the local bank at an interest rate of 9% per year. What is their monthly payment?

After living five years in this house, they were transferred by their employer to another division in a different state, and they wanted to sell the house.

a. How much of the principal is left at the end of the fifth year?

b. If their MARR is 10%, what should be their minimum asking price for the house?

5- A successful physician has invested $800,000 cash in a rental apartment house. If he has a MARR of 10%, how much should he charge for rent per month to recover his investment in 10 years?

6-You have the option of choosing between the following two projects.

a. Initial investment $700K, annual income $400K

b. Initial investment $ 1,600K, annual income $600K

The life of project a is five years and that of project b is ten years. If you have a MARR of 30%, which one of the projects should you accept?

7- Mr. Goodman, a friend of yours, is asked to invest in the following project:

installation and operation of a facility with a life span of five years. The initial investment is

$90M. It will have a net profit of $25M/Yr the first two years and of $30M/Yr in years 3,4,

and 5. At the end of year 5, it has to be disposed of at a cost of $10M with no resale value.

He also has the option of investing the same money in a project that will bring him $29M per

year. If he has the money and his opportunity cost of money is 10% (I=10%), which proposal

do you advise him to accept? Why? Explain.

8- A developer is given the following two options for the purchase of a property:

a.   Pay $100,000

b.   Pay $30,000 at the end of each year, starting one year after purchase, for the next five

years. Which option should he take?

9- In engineering economic analysis of projects, when using Net Present Worth or

Benefit/Cost ratio we have to equalize the lives of the projects. This equalization is not

necessary when using EUAW methods. Why? Can you show this using a cash flow

diagram?

CHAPTER 5

Rate of Return

RATE OF RETURN

This is yet another useful method for comparing the financial advantages of alternative systems using the cash flow diagram. We calculate that specific rate of interest for the system that makes the net present value equal to zero. This rate is called the rate of return (ROR) and is denoted by $i*$. If this rate is higher than the minimum rate that satisfies the investor or the project manager, then the project is acceptable. This minimum rate is called the Minimum Acceptable Rate of Return (MARR). There is no mathematical formula for calculating MARR. This has to be done by trial and error. Fortunately, there are computer programs that make this calculation simple and fast. Most of the spreadsheets on the market, such as Quattro Pro, Excel, etc., have provisions for calculating the rate of return.

Example 5.1

Let us see what the ROR is for the problem of Example 2.1 using the netted cash flow diagram.

Assuming an unknown interest rate $i*$, we can write the NPW as

$$NPW = -24000 + 4600 \ (P/A, \ i*, \ 3) + 19600 \ (P/F, \ i*, \ 4)$$

By definition ROR is the interest rate that makes NPW = 0.

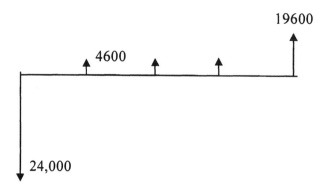

To determine ROR, we have to try several values for $i*$ and see which one makes NPW = 0.

If the value of ROR is higher than the MARR, then the project is good.

Calculation of $i*$ from the above equation is not easy, and we have to use trial and error. To

do this, we have to arbitrarily pick several (at least three) values for $i*$ and calculate the

corresponding value for NPW from the above equation. We can then graph these values and

find the value of $i*$ from the point that NPW value intersects the interest rate axis.

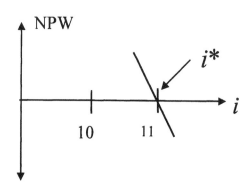

This is a cumbersome job; fortunately, there are financial calculators that perform this. Almost all spreadsheet programs can perform this calculation as well. In the case of this particular example, the answer is

$$ROR = 11.23\%$$

Use of the spreadsheet will give us the same answer.

	A	B	C	D	E	F	G	H
1	Example 5.1							
2			MARR=	10%				
3		Year	0	1	2	3	4	
4			-24000	4600	4600	4600	19600	
5								
6		For Excel use =IRR(C5..G5) at G7					11.23%	
7		For Quattro Pro use @IRR(10%C5:G5) at G					11.23%	

MULTIPLE ALTERNATIVES AND INCREMENTAL ANALYSIS

As was in the case of other analysis methods discussed, very often the decision is to choose the best of two or more alternatives. At first glance, it seems logical that the alternative with the highest rate of return is preferred. This is true only in the specific case where the initial investment is the same for all available alternatives. In cases where the initial investment is not the same, the alternatives are

A: To invest in the lower initial investment case and invest the rest of the capital where the investor can get his MARR.

B: To invest in the program with the higher initial investment.

To solve this problem, we perform what is called "incremental analysis". The procedure is

Step 1:

Set up the cash flow of all alternatives in ascending order of initial investment.

Step 2:

Discard all alternatives that have an ROR less than the MARR. This step can be ignored but performing it will save time on the rest of the steps.

Step 3:

Construct the cash flow of the difference of alternatives two by two, starting from the two with the lowest ROR. Always subtract the one with lowest initial investment from the one with the higher initial investment. A check on this step is that the cash flow at time zero of the differential should always be negative.

If the ROR of the differential is higher than the MARR, the alternative with the higher initial investment is preferred, otherwise the one with the lower initial investment is preferred. Next we compare the preferred alternative with the next alternative on the list in the same manner.

Example 5.2

An investor is offered two investment opportunities. Project A is an investment in frozen yogurt equipment that requires an initial investment of $40,000 with a life of three years. Its annual operating costs and annual incomes are presented in table A. The equipment can be sold at the end of year 3 at a resale value of $5,000.

Table A

Year	1	2	3
Income	25000	33000	45000
Expense	15000	18000	20000

The second opportunity is purchase of printing equipment with an initial investment of $200,000. Annual operating costs and annual incomes are presented in table B. The equipment can be sold at the end of year 5 at a resale value of $20,000.

Table B

Year	1	2	3	4	5
Income	110000	135000	170000	210000	230000
Expense	100000	110000	115000	120000	130000

If the MARR for this investor is 10%, which project should he invest in? Ignore tax and depreciation.

Step 1:

Draw the cash flow diagram of the frozen yogurt project, and check it against the null alternative.

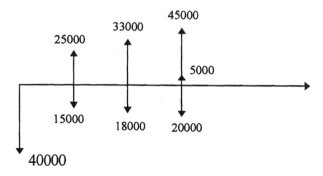

The net cash flow is

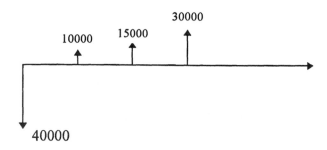

NPW = -40000 +10000(P/F, i^*,1) +15000(P/F, i^*,2) + 30000(P/F, i^*,3)

By trial and error, $i^* = \underline{14.7\%}$　　　　$\underline{i^* > 10\%}$

Therefore, this alternative compared to the null alternative is acceptable

<u>Step 2</u>:

　　Draw the cash flow diagram of the print shop proposal.

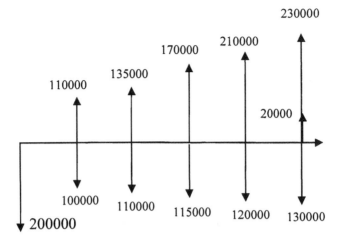

The net cash flow diagram in this case is

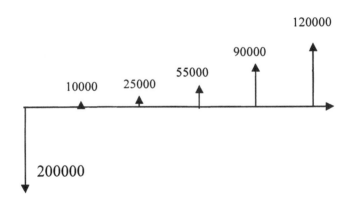

Calculating ROR as we did in the case of project A we obtain ROR=11%.

At first look, we are inclined to state that we should choose project A, which means we invest

$40,000 in project A and invest the rest where we can get the required MARR of 10%.

But let us follow the rules and do the incremental analysis. We will see that our hunch is not

correct.

<u>Step 3:</u>

We have to use the incremental cash flow diagram, i.e.,. B-A. <u>In calculating ROR, we do not have to equalize the lives.</u> That makes it easier. The incremental cash flow is

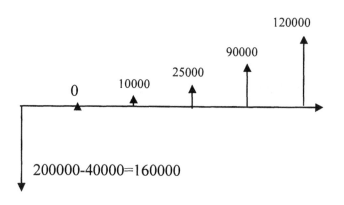

<u>Step 4:</u>

The ROR for this incremental cash flow is calculated as follows:

NPW= -160000 +0(P/F, $i*$,1) + 10000(P/F, $i*$,2) + 25000(P/F, $i*$,3) +90000(P/F, $i*$,4)

+120000(P/F, $i*$,5) = 0

By trial and error, $i*$ = 10.5. Therefore, <u>$i*$>10%</u>; hence, project B is preferred.

This means it is better to invest all the available money ($200,000) in project B at 10.5% rather than invest $40,000 in project A and the rest at MARR of 10%.

Another way of testing is to substitute the value of MARR for $i*$ in the equation obtained for the NPW of the incremental analysis of the two projects. If the resulting NPW is positive, project B is preferred.

An interested reader can follow the hunch and calculate the return of the combination of choosing project A and investing the difference between the initial payment of project A and B at the MARR and see that the initial hunch is not correct. The spreadsheet for Example 5.2 is shown below. The discrepancy is due to rounding to different decimal points.

	A	B	C	D	E	F	G	H
1	**Example 5.2**							
2			MARR=	10%				
3								
4	Project A:							
5		Year	0	1	2	3	4	5
6		Income	0	25,000	33,000	50,000		
7		Expens	40,000	15,000	18,000	20,000		
8		Net	(40,000)	10,000	15,000	30,000		
9								
10			For Excel use =IRR(C8..F5) at G7				IRR=	14.70%
11			For Quattro Pro use @IRR(10%,C8:F8) at H8				IRR=	14.70%
12	Project B							
13		Year	0	1	2	3	4	5
14		Income	0	110,000	135,000	170,000	210,000	250,000
15		Expens	200,000	100,000	110,000	115,000	120,000	130,000
16		Net	(200,000)	10,000	25,000	55,000	90,000	120,000
17			For Excel use =IRR(C16..H16) at H17				IRR=	11.00%
18			For Quattro Pro use @IRR(10%,C16:H16) at H18				IRR=	11.00%
19								
20	Incremental: Project B - Project A							
21			(160,000)	0	10,000	25,000	90,000	120,000
22								
23			For Excel use =IRR(C21..H21) at 23				IRR=	10.49%
24			For Quattro Pro use @IRR(10%,C21:H21) atH 24				IRR=	10.49%
25								

MULTIPLE RATES OF RETURN

In some cases, there is more than one value of the ROR that makes NPW = 0. This can be detected when the net cash flow diagram goes from cost to benefit more than one time. In fact, according to Descartes' rule of signs, the number of the RORs equals the number of changes from cost to benefit or vice versa. In these cases, it is advisable to perform the analysis using the NPW method.

PROBLEMS

1-The cash flow of an investment is shown below. What is the rate of return for this

investment (i.e., IRR or ROR)? Calculate the rate to two digits.

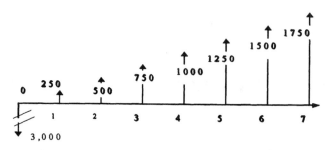

2- A local corporation purchased a system for $1 million. The net income from operating this

system is $300,000 per year. Assuming a life of five years and no salvage value, what is the

rate of return of this system?

3- Darbol Corporation received two investment proposals. The estimate of the financial

situation of each proposal is presented in the following table. Darbol also has the choice of

investing the capital and receiving an interest rate of 10% annually.

Using the ROR method, perform the financial analysis and make your recommendation

as to which of the proposals, if any, they should accept.

Data	Proposal A	Proposal B
Initial Cost	$50,000	$140,000
Salvage Value	$20,000	$30,000

Annual Benefit	$30,000	$40,000
Annual Cost	$15,000	$10,000
Life	3 years	6 years

4-An investor is about to make an investment of $45,000 in equipment that will bring a net annual income of $20,000. The equipment will last for only three years, has no resale value, and removal cost is negligible. An engineer suggests an improvement in the equipment that will cost $35,000 and will extend its life to six years but will not change the annual income. The equipment will have no resale value after the six-year life; the removal cost is again negligible. The investor has MARR of 10%. Use the rate of return method to make a recommendation to him. Explain your reasons for recommendation.

5- The following two projects are recommended for investment:

    a. Initial investment $800K, annual income $400K

    b. Initial investment $1,500K, annual income $700K

The lives of both projects are five years. The investor has a MARR of 30%. Using the ROR method, make a recommendation.

6- Mr. Friendly, a friend of yours, is asked to invest in the following project:

Installation and operation of a facility with a life span of five years. The initial investment is $90M. It will have a net profit of $25M/Yr the first two years and $30M/Yr in years 3,4, and 5. At the end of year 5, it has to be disposed of at a cost of $10M with no resale value.

a. If he has the money and his opportunity cost of money is 10% (i=10%), would you advise him to invest or not? Yes? No? Why? Explain.

b. He can only put down $33M, and his bank will extend him a loan for the rest of the initial investment to be used on this project at 15% in a way that he has to pay it back in equal installments at the end of each of the five years with no collateral. Should he take the loan and invest? Yes? No? Why? Explain.

7- Production equipment is bought at an initial price of $10,000. The annual operation and maintenance cost is $100. The salvage value at the end of the 15-year life is $500. If the equipment brings in an income of $1,100 per month, what is the rate of return for this project?

8- Members of the board at ACE Corporation received three proposals for a machine they may want to purchase. They also have a choice of investing their capital and receiving an interest rate of 15% annually. Using the following data, what is their most economical course of action? Use a 10-year life span.

Data	Machine A	Machine B	Machine C
Initial Cost	$180,000	$235,000	$200,000
Salvage Value	$38,300	$44,800	$14,400
Annual Benefit	$75,300	$89,000	$68,000
Annual Cost	$21,000	$21,000	$12,000

CHAPTER 6

Benefit-Cost Ratio and Payback Methods

BENEFIT-COST RATIO

Another method of assessing the viability of a system or comparing several systems is to calculate the net present value of the costs and the benefits and obtain the benefit-cost ratio (B/C). If this ratio is greater than one, then the project is profitable.

Example 6.1

A car leasing company buys a car from a wholesaler for $24,000 and leases it to a customer for four years at $5,000 per year. Since the maintenance is not included in the lease, the leasing company has to spend $400 per year in servicing the car. At the end of the four years, the leasing company takes back the car and sells it to a secondhand car dealer for $ 15,000. For the moment, in constructing the cash flow diagram, we will not consider tax, inflation, and depreciation.

Step 1:

The cash flow diagram is

67

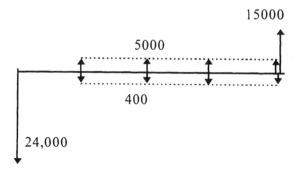

Step 2:

Calculate the net present value of the benefits and the costs.

Net Present Benefit = NPB = 5000 (P/A,10,4) + 15000 (P/F,10,4)

$$= 5000 \ (3.17) + 15000 \ (0.6830) = \underline{26,095}$$

Net Present Cost = NPC = 24,000 + 400 (P/A,10,4) = <u>25,268</u>

Hence, B/C = 26095/25268 = 1.033   The project is acceptable.

The spreadsheet solution of this example is exactly as in the case of the NPW (Chapter 2).

EQUALIZING LIVES AND INCREMENTAL ANALYSIS

If two or more projects have to be compared with the B/C method, then the lives should be equalized as in the case of the present worth method. We also have to calculate the B/C of their differential cash flows in pairs and perform incremental analysis <u>to ensure that the extra initial cost justifies the extra benefit</u>. We can see that the B/C ratio is a difficult comparison method when more than one alternative

is considered.

Example 6.2

Suppose the leasing company of Example 6.1 has to choose between the

following two projects:

1.   Lease the car exactly as in Example 6.1.

2.   Buy a car at $25,000, lease it for two years at $10,000 per year with

no maintenance cost and sell it for $18,000 at the end of two years.

Assuming an interest rate of 10%, which project should we choose?

In this problem, $n_1=4$ and $n_2=2$; therefore, the least common multiplier of n_1 and n_2

is equal to 4. That means that project 1 continues for 2 years after project 2 ends.

Step 1:

Start with the project that has the lowest initial investment. Perform the

B/C analysis to see if the project is acceptable (i.e., B/C >1).

This is checking against the null alternative. We know from Example 6.1 that

B/C= 1.033, hence the project is economically acceptable.

Step 2:

Construct the netted cash flow diagram of the two projects at equalized

lives. In this case, the common denominator of the lives of the two projects is four

years; therefore, the cash flow diagram is constructed for four years.

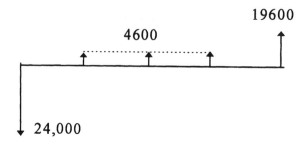

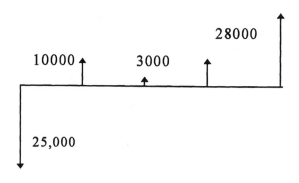

Step 3:

Construct the incremental cash flow diagram.

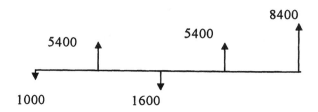

<u>Step 4:</u>

Perform the incremental analysis, i.e., calculate the B/C of the differential.

NPB = 5400 (P/F,10,1) + 5400 (P/F,10,3) + 8400 (P/F,10,4)

   = 5400 (0.9091) + 5400 (0.7513) + 8400 (0.6830) = 14,703.36

NPC = 1000 + 1600 (P/F,10,2) = 1000+1600 (0.8264) = 2322.24

Therefore,

       B/C = 6.33   Hence, project 2 is preferred.

If a choice has to be made between several projects, rank them by increasing initial investment. Compare them two by two in the same manner as we did for the above two projects, pick the best one, and compare it with the next in the ranking.

PAYBACK

A simple crude method for getting a quick evaluation of the alternatives is to calculate how long it takes to recover the initial investment. The time in any unit that it takes to recover the initial investment is called the payback period. In this method, we first construct the net cash flow diagram and then by simple arithmetic calculation add the benefits and the cost year by year until the total equals the initial investment. It is obvious that the payback period neglects the time value of money and is only accurate when the interest rate is zero. Even with this shortcoming, many analysts consider this method to be a useful quick and dirty way of comparison.

Example 6.3

A \$100,000 investment is made into a project. The net benefit of this project is \$15,000 per year. What is the payback period?

Solution:

We first construct the net cash flow diagram.

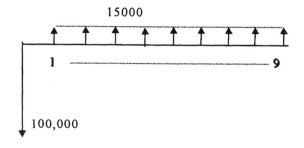

At year 6 the total income is   6 * 15000 = 90000

At year 7 the total income is   7 * 15000 = 105000

Hence sometime between year six and seven or to be precise in six years and eight months, the initial investment is recovered. The payback period is then 80 months.

MULTIPLE ALTERNATIVES

In the case of multiple alternatives the one with the least payback time is the preferred alternative.

PROBLEMS

1-The cash flow of an investment is shown below. If the MARR for the investor is 15%, What is the B/C ratio?

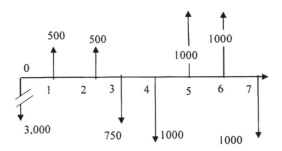

2- An investment requires an initial payment of $1,000. The table below indicates the estimates of income and expense for this investment. What is the B/C for MARR of 10%? Is this a good investment?

Year	1	2	3	4	5
Income	70000	80000	90000	100000	110000
Expense	50000	60000	70000	80000	100000

3-You are asked to investigate two investment proposals with the estimates of income and expense shown below. There is also the possibility of investing the capital and receiving an interest rate of 15 % annually. What would be the best choice for your client? Use both B/C ratio and payback methods.

Proposal	Initial Cost	Annual Benefit	Annual Cost	Salvage Value	Life
A	75,000	45,000	22,000	30,000	2 Year
B	2,100,000	60,000	15,000	45,000	4 Years

CHAPTER 7

Inflation

INFLATION AND PURCHASING POWER

Inflation reflects the buying power of money. Suppose you buy a unit of a certain commodity, e.g., a pound of sugar for $X, today and at some future time its price changes to $Y. If Y>X then the buying power of your money has been reduced, that is, you have to pay more money to buy the same commodity. Put differently, one dollar buys less of that special commodity, e.g., sugar, as time passes. For any period,

$$InflationRate = f = \frac{Y - X}{X}$$

If Y<X then the change is called a deflation. Usually, although not always, the buying power of money decreases with time. In other words, the same item has a higher dollar value as time progresses.

Obviously, the rates of inflation or deflation depend on the particular item under consideration. Economists use a basket of items for consideration and use the aggregate, weighted rate of inflation of the items in the basket as the general inflation rate of the economy. The time period is usually one year, and the basket is made up of the commodities that most affect the life of citizens. The index used most often to reflect the inflation is the consumer price index (CPI). The basket used for this index is made up of consumer goods.

75

For a constant rate of inflation f, the buying power referred to the base year can be calculated from equation 7.1a.

$$S_0 = S_n(1+f)^{-n} \qquad\qquad (7.1a)$$

Where

$\quad$ S_n = Dollar value after n years

$\quad$ S_0 = Equivalent dollars at the base year

$\quad$ f = Rate of inflation

The values calculated from equation 7.1a are called the constant dollar values.

We see that equation 7.1a is exactly the same as equation 2.2a with f replacing i. We can therefore write expression 7.1b which is the counterpart of expression 2.2b.

$$S_0 = S_n(P/F,i,n) \qquad\qquad (7.1b)$$

In effect, inflation rate and interest rate work the same way, the former influences the buying power of money, and the latter influences the earning power of money.

INFLATION AND INTEREST RATE COMBINED

$\qquad$ Investors know that they should earn enough on their investment to compensate for inflation. Therefore, they adjust their minimum acceptable rate of return (MARR) to reflect the expected inflation rate during the investment period. If the rate of inflation, f, is constant or can be assumed to be constant during the life of the project, then we can use equation 7.2a or expression 7.2b to compensate for the effect of inflation. The result of the financial analysis would be in constant dollar referred to the year of the start of project.

$$u = i + f + if \qquad (7.2a)$$

$$i = \frac{u - f}{1 + f} \qquad (7.2b)$$

u = prevalent interest rate (apparent interest rate)

i = effective interest rate with inflation accounted for (real interest rate)

f = inflation rate for the item under consideration

This means that although the bank states that it is giving you an interest of u per a given period, you are in effect receiving an interest rate of i that is less than u. Conversely, if you want to receive an effective interest rate of i and inflation is expected to be f, then you should shop around for an interest of u calculated by the transposition of equation 7.2b. That is

$$u = i + f + if$$

Example 7.1

The salary of an engineer is increased every year according to the following table. The inflation rate during these years has been a constant 3%. Calculate the 1991 constant dollar value of these salaries.

Year	Yearly salary	Constant $ values
1991	30,000	30,000
1992	33,000	32,040
1993	36,000	33,934
1994	40,000	36,604

Solution:

Equations 7.1a or 7.1b were used to calculate the 1991 constant dollar values of the stated

salaries.

S1 = 33000 (0.9709) = 32040

S2 = 36000 (0.9426) = 33934

S3 = 40000 (0.9151) = 36604

The third column in Example 7.1 is the constant dollar values of salaries referred to 1990. We

can see that considering the buying power of money, the salary increase is not as much as it

seems at first look.

If the rate of inflation is not constant over several years, then the constant dollar value should

be calculated year by year using annual values.

INFLATION ADJUSTED DISCOUNT RATE

As you can see from the above example, the present value (year one value) of each

year's payment is reduced due to inflation. We also know that the present value of any year

payment is reduced due to the time value of money as represented by the interest rate. The

combination of these two factors reduces the present value of any year payments according to

a factor denoted by i/f and called the inflation adjusted discount rate.

$$i/f = (1+i)(1+f)-1$$

$$= i+f+if \qquad (7.3)$$

Equation 7.3 is the same as equation 7.2a.

Example 7.2

A developer is given the following two options for the purchase of a property:

a. Pay $120,000

b. Pay $32,000 at the end of each year, starting one year after purchase and continuing

for the next six years.

If the inflation rate is assumed to be 4.4% per year and the interest rate 15%, which option

should he take? Ignore tax.

Solution:

According to equation 7.3, the inflation-adjusted discount rate is

$$0.44+0.15+(0.44)(0.15) = 0.2006 \text{ or } 20\%$$

Therefore, the NPW of option b is

$$NPW = 32000(P/A, 20, 6) = 32000 (3.326) = 106,432$$

Option b is preferred.

DEFLATION

In some economic circumstances, the price of certain goods falls as the years go by. This most occurs in the case of new product, especially when incorporation of new technology is involved. A case in point is the case of personal computers where the prices fall almost monthly. If we consider capability per dollar, the reduction is very noticeable. The communication industry is also experiencing a price reduction. The effect of this reduction in price is called deflation. Deflation can be considered a negative inflation; therefore, the same rules discussed above can be applied to it.

GENERAL INFLATION INDICES

As mentioned previously, the inflation rate of several products combined with their weighted effect on consumption of the general population of a society is the inflation index of that society. Several indices are used to measure the effect of inflation. An example is the consumer price index (CPI).

To calculate the CPI they assume an average citizen. Then they figure out statistically how much of any commodity (meat, cooking oil, vegetable, gas, electricity, etc.) this average citizen consumes per year. They multiply these by their respective unit price and add it all together and this will give an indication of the dollar value of the average citizen. Let us call this $V(n)$. When they do this for every commodity, they can obtain the rate of inflation between any two years as the ratio of $V(n_2)/Vn_1$. Normally a specific year is chosen as the base year and the inflation rate is indexed relative to the base year.

As an exercise, the reader can choose a set of commodities of his concern and calculate the inflation rate that affects his lifestyle.

PROBLEMS

1- A developer is given the following two options for the purchase of a property:

a.   Pay $100,000

b.   Pay $30,000 at the end of each year, starting one year after purchase, for the next five

years. If the inflation rate is assumed to be 5% per year and the interest rate 10%,

which option should he take? Do not consider tax effects.

2- An investor is offered a project that needs an initial investment of $240,000. The

investment produces a net profit of $20,000 in the first year increasing by $10,000 each year

for five years after that (a total project duration of six years), with a final payment of $15,000

at the termination (end of the sixth year). If the inflation rate is assumed to be a constant for

the next 10 years at 4.545% per year, and the prevailing market interest rate is estimated to be

15% per year for the duration of the project, should he invest in this project? Ignore tax.

3- An investor is offered the opportunity of investing $100,000 in a six year project that will

have a net income of $25,000 per year.

a.   If the inflation rate is a constant 4% per year and his after inflation MARR is

10%, should he invest in this project?

b.   If he is looking to invest his $100,000 where he can get an inflation adjusted

rate (i.e., net, after inflation is taken into account) of 12%, what minimum rate

of interest should he shop for?

4-An investor has two opportunities. An investment project in country A involving an initial investment of $100,000 and an income of $30,000 for six years, or, an investment project involving an initial investment of $150,000 in the US with an annual income of $60,000 for three years. By how many "inflation adjusted" US dollars would this investor be better or worse off, if he invests in the country A project compared to the US project. The prevailing rate of interest in both countries is 15%. Assume that

a. The rate of inflation in the US is constant at 5.5%, and the inflation rate in country A is constant at 4.545 per year.

b. The rate of exchange between the US and country A currency remains the same. Disregard social, stability, economic, and other conditions.

CHAPTER 8

Tax and Depreciation

EFFECTS OF INCOME TAX

So far we have ignored the effect of taxation on the financial analysis. The government, for reasons beyond the scope of this book, levies tax on the profits of a project. In the years that the project loses money, this loss can be used to recover an equivalent part of the profits of other projects owned by the same organization. It is a practice in all financial analysis calculations to assume that there is always a project to which this loss can be applied to. Depreciation, as will be discussed in this chapter, is considered a legitimate expense while in reality it is a virtual expense and is used to recover the initial cost.

Example 8.1

The yearly net before tax income (IBT) of a project is shown in the second column of the following table. Tax is assumed to be 30% of the IBT. Calculate the before- and after-tax NPW of the project. Table 8.1 shows the resultant income after tax (IAT).

Year	IBT	Tax = 0.30	IAT = IBT - Tax
1	+ 100,000	+ 30,000	+ 70,000
2	- 30,000	- 10,000	- 30,000 - (-10,000) = -20,000
3	+ 200,000	+ 60,000	+ 140,000

Table 8.1

83

Before-tax cash flow:

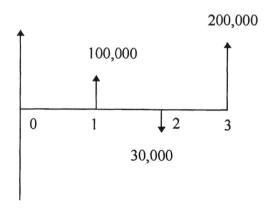

NPW = 100,000 (P/F,10,1) - 30,000 (P/F,10,2) + 200,000 (P/F,10,3) = 216,378

Cash flow after tax:

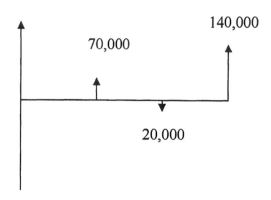

NPW = 70,000 (P/F,10,1) - 20,000 (P/F,10,2) + 140,000 (P/F,10,3) = 152,291

In a similar manner, an after-tax NFW, ROR, and EUAW can be calculated. We could also use the spreadsheet and obtain the same result.

DEPRECIATION, BOOK VALUE, AND CAPITAL GAIN

When a corporation purchases equipment, a facility, etc., that is used more than a year, it has to recover the cost over a number of years. This is called depreciating, and the process is depreciation. The annual depreciation from the depreciation formulae or tables is deducted each year from the income in the same manner as regular cost. The remaining cost of the purchase is called the book value.

Book Value at year n = Initial Cost - Total Depreciation up to year n

At any time, if the purchased equipment, property, etc., is sold, the difference between the sale price and the book value is called capital gain or loss and is considered an income or loss and is treated accordingly for tax purposes. This is demonstrated in Example 8.2.

The process of depreciation is governed by depreciation rules (some set by the IRS). We will discuss four depreciation procedures or schedules here.

STRAIGHT-LINE DEPRECIATION

In this method the cost is spread uniformly over a certain number of years N called the life. The definition of the cost for this purpose is the initial cost P minus the estimated resale

value S at the end of the project.

$$Depreciation/Year = \frac{P-S}{N} \qquad (8.1)$$

Example 8.2

What is the yearly depreciation and the book value for a truck with an initial cost of

$150,000, an assumed life of five years, and an expected resale value of $50,000?

$$D = \frac{150,000 - 50,000}{5} = 20,000$$

Year	Depreciation	Book Value
1	20,000	130,000
2	20,000	110,000
3	20,000	90,000
4	20,000	70,000
5	20,000	50,000

The resale price of the truck in the calculation was estimated; we do not know what the actual

sale price would be. If the truck is sold at the end of year 5 for $70,000, then

$$Capital\ Gain = 70,000 - 50,000 = 20,000$$

If it is sold for $20,000, then

$$Capital\ Loss = 20,000 - 50,000 = -30,000$$

Capital gain or loss is treated as ordinary benefit or cost. Some years ago only part of the gain was treated as ordinary gain. This law may again come into effect in the future.

DOUBLE DECLINING BALANCE (DDB) DEPRECIATION

In this method, the salvage value is estimated but is not considered in the calculation. Depreciation for any year n is calculated from equation 8.2. This is done until the end of the estimated life or the year for which the book value is less than or equal to the estimated resale value, whichever comes first. After this year, we have to stop depreciating. In other words, the book value should never drop below the estimated resale. We must remember that the resale value is just an estimate that can be set at any value so as to have the effect we want on the depreciation schedule.

$$D_n = \frac{2}{N} (P - \sum_1^n D) \qquad (8.2)$$

Where $\sum_1^n D$ is the sum of deprecations to year n.

Example 8.3

Use the same values of Example 8.2 and calculate depreciation using double declining balance.

Year 1 $D = 2/5 (150,000) = 60,000$ Book Value 90,000

Year 2 $D = 2/5 (150,000 - 60,000) = 36,000$ Book Value 54,000

Year 3 $D = 2/5 (150,000 - (60,000 + 36,000)) = 21,600$ Book Value 32,400

This book value is less than the estimated resale value of $50,000. Therefore, in the third year we can only depreciate a dollar value that makes the book value equal to $50,000.

Hence, depreciation at year 3 is $54000 - 50000 = 4000$

The capital gain or loss is calculated the same way as in Example 8.2. The Quattro Pro and Excel spreadsheet calculations of DDB depreciation are presented below. Quattro Pro calculates the depreciation for any year, but Excel gives the total depreciation up to any year.

Quattro Pro:

	A	B	C	D	E	F
1	Example 8.3:					
2						
3	Cost =	150000				
4	Resale =	50000				
5	Number of year =	5				
6	Use @DDB (Cost,,Salvage,Life,year)on column F					
7		Year 1	@DDB(B3,B4,C5,1)			60000
8		Year 2	@DDB(B3,B4,C5,2)			36000
9		Year 3	@DDB(B3,B4,C5,3)			4000
10		Year 4	@DDB(B3,B4,C5,4)			0
11		Year 5	@DDB(B3,B4,C5,5)			0

Excel:

E11		=					
	A	B	C	D	E	F	G
1	Example 8.3						
2							
3		Cost	150000				
4		Resale	50000				
5		Life	5			Depreciation	
6		Use =VDB(C3,C$,C%,0,1) at E6				Up to Year 1= 60,000	
7		Use =VDB(C3,C$,C%,0,2) at E6				Up to Year 2= 96,000	
8		Use =VDB(C3,C$,C%,0,3) at E6				Up to year 3= 100,000	
9		Use =VDB(C3,C$,C%,0,4) at E6				Up to year 4= 100,000	
10							

MODIFIED ACCELERATED COST RECOVERY SYSTEM (MACRS) DEPRECIATION

This is the depreciation method mostly used in industry. It is based on IRS publication 534. This publication classifies properties into several classes and states the allowable life for each. It also has a table that gives the percentage of the total initial cost that can be deducted as depreciation for each year of the allowable life. The depreciation schedule for four classes of properties is presented in table 8.2. The IRS publication mentioned above gives more detail about which equipment is in which class and provides the depreciation schedule.

Recovery Year	3 Year Class	5 Year Class	7 Year Class	10 Year Class
1	33.33%	20%	14.29%	10%
2	44.45%	32%	24.49%	18%
3	14.81% *	19.2%	17.49%	14.4%
4	7.41%	11.52% *	12.49	11.52%
5		11.52%	8.93% *	9.22%
6		5.76%	8.92%	7.37%
7			8.93%	6.55% *
8			4.46%	6.55%
9				6.55%
10				6.55%
11				3.28%

Table 8.2

The calculation is based on the double declining balance with conversion to the straight-line method at the years denoted by *. This publication makes the job of accountants easy as far as depreciation is concerned.

Example 8.4

Automobiles and trucks fall into the five-year category. Therefore, if the total initial

cost of a truck is $150,000, the depreciation and book value of this property from the above

table is

Year	Depreciation	Book Value
1	30,000	120,000
2	48,000	72,000
3	28,800	43,200
4	17,280	25,920
5	17,280	8,640
6	8,640	0

SUM OF YEARS DIGIT (SOYD) DEPRECIATION

In this method, depreciation for each year is calculated using a number called the sum

of years digit (SOYD). This number is calculated from equation 8.3.

$$SOYD = N/2 \ (N+1) \qquad\qquad (8.3)$$

N = Estimated useful life of the system

Depreciation for each year is then calculated from equation 8.4.

$$\frac{RL}{SOYD}(P-S) \qquad\qquad (8.4)$$

Where

RL = Remaining life at any year, at the beginning of the year

P = Purchase price

S = Estimated resale value

Example 8.5

Calculate the annual depreciation values for the case of Example 8.2 using the

SOYD method.

SOYD = 5/2 (5+1) = 15

Depreciation for year 1 = 5/15 (150000-50000) = 33333.3

Depreciation for year 2 = 4/15 (150000-50000) = 26666.7

Depreciation for year 3 = 3/15 (150000-50000) = 20000

Depreciation for year 4 = 2/15 (150000-50000) = 13333.3

Depreciation for year 5 = 1/15 (150000-50000) = 6666.7

CHANGE OF DEPRECIATION METHODS

As can be seen from the above examples, the rate of change of depreciation, i.e., how fast an asset is depreciated, is different for each type of depreciation procedure. Therefore, accountants for reasons beyond the scope of this book use different depreciation methods. They sometimes change from DDB to straight-line depreciation when it is appropriate. This is also done by the IRS as can be seen in the table of MACRS.

AMORTIZATION AND DEPLETION

An organization may own or acquire non-tangible assets. These are assets such as patent rights, goodwill, and right of way. They are amortized instead of depreciated.

As far as financial analysis is concerned, the effect is the same as depreciation. The procedure

is also the same, and the same methods could be used.

Depletion is also similar to depreciation. It applies to assets such as mines, oil wells, etc. The

annual portion of the acquisition cost allocated as expense depends on the percentage of the

total estimated quantity that is extracted in any particular year.

FINANCIAL ANALYSIS WITH TAX AND DEPRECIATION

 In financial analysis, depreciation is treated as an expense, although no actual payment

is made to anyone in that particular year. It therefore reduces the taxable income and increases

the income after tax. The combined effect of depreciation and tax can be illustrated best by an

example.

Example 8.6

 Independent Speedy Printing of Fairfax has the following expected annual income and

expense:

Year	1	2	3	4	5
Income	$100,000	$120,000	$150,000	$200,000	$180,000
Expense	$130,000	$120,000	$130,000	$180,000	$160,000

In the beginning of year 1 it buys a copying machine for $15,000. The company is using the straight-line depreciation method and life of five years for the machine and assumes no resale

value at the end of its life. The company's tax rate is 32%. At the end of five years, they sell

the machine for $3,000 and the printing business for $200,000. Assume that the company has

other income to deduct the losses from. At an interest rate of 10% what are the NPW, NFW,

EUAW, and ROR of this project?

Solution:

Operating Profit = OP = Income - Expense

Depreciation = (15,000- 0) / 5 =3,000 per year

Taxable Income (IBT) = Operating Profit - Depreciation

Tax = 32% of IBT; therefore, income after tax, IAT = OP-TAX

Financial analysis is facilitated by construction of the following table.

Year	Total Income 1	Total Expense 2	Operating Profit 3=1-2	Depr. 4	BK.Val 5	Taxable Income 6=3-4	Tax 7=6*TR	IAT 8=3-7
0		15000						
1	100,000	130,000	-30,000	3,000	12000	-33,000	-10,560	-19,440
2	120,000	120,000	00,000	3,000	9000	-3,000	-960	960
3	150,000	130,000	20,000	3,000	6000	17,000	5,440	14,560
4	200,000	180,000	20,000	3,000	3000	17,000	5,440	14,560
5	180,000	160,000	20,000	3,000	0	17,000	5,440	14,560
Bus. Sale	200,000					200,000	64,000	136,000
Eqp.Resale	3,000					3000	960	2,040
Total	383,000							152,600

Note that at year 4 we have two extra incomes. The sale of the business brings in $200,000. Since the book value at the end of year 5 is zero, the capital gain is $3,000 - 0 = 3,000$. The taxes on these two incomes are at 32% and are shown separately in column 6. The cash flow diagram is

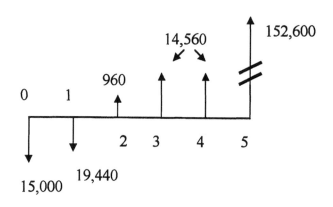

$NPW = -15000 - 19440 \ (P/F,10,1) + 960 \ (P/F,10,2) + 14560 \ (P/F,10,3) + 14560 \ (P/F,10,4) +$

$152,600 \ (P/F,10,5) = 83,765.64$

$EUAW = NPW \ (A/P,10,5) = 21,992.16$

$NFW = -15,000 \ (F/P,10,5) - 19440 \ (F/P,10,4) + 960 \ (F/P,10,3) + 14550 \ (F/P,10,2)$

$+ 14550 \ (F/P,10,1) + 152600 = 134,903.88$

<u>ROR by trial and error $= 48.06\%$</u>

A suitable semi-generic spreadsheet can be constructed to solve this kind of problem as shown below. As usual, the difference between the outcome from the spreadsheet and the manual calculation is due to rounding of numbers. The spreadsheet shown is actually the representation of the top line of the generic table constructed for comprehensive Example 8.6 and can be used for all problems of this kind with minimum adjustment.

Ezzample 8.6

Cost	15,000		Resale					
Years	5		Estimate	0	Iterest	10%		
Good wil	200,000		Actual	3,000	Tax Rate	32%		
		Depreciation SL	3,000					

Year	Income	Expense	OP	Depr.	Bk. Value	IBT	Tax	IAT
0	0	15,000	-15,000					
1	100,000	130,000	-30,000	3,000	12,000	-33,000	-10560	-19,440
2	120,000	120,000	0	3,000	9,000	-3,000	-960	960
3	150,000	130,000	20,000	3,000	6,000	17,000	5440	14,560
4	200,000	180,000	20,000	3,000	3,000	17,000	5440	14,560
5	180,000	160,000	20,000	3,000	0	17,000	5440	14,560
Cap. Gain	203,000		203,000			203,000	64960	138,040

Capital Gain = Good wil + actual resale and is taxed the same as regular income

Year	0	1	2	3	4	5
Cash Fl.	-19,440	-19,440	960	14,560	14,560	152,600

* Net cash flow in the last year is the sum of the after tax vlaues of the income from The operations and that of the good will.

* Now you can use the NPW and IRR spreadsheet formulae as in chapters 2 to 5

| ROR= | 49.67% | NPV= | $83,757 |

PROBLEMS

1- The initial cost of a heavy-duty truck is $140,000 It is assumed that its resale value after five years is $30,000. Calculate its depreciation schedule using:

a- Straight-line method

b- Double declining balance method

c- Sum of the years digit method.

d- MACRS method. Note- the truck is a five-year class property for the MACRS method.

2- A service line of business has an initial non-depreciable investment of $500,000 and an annual net income of $100,000. It operates for 10 years with no resale value. If their tax rate is 30%, what is the rate of return of this investment at MARR=10%?

3-The initial cost of equipment is $1,000,000. Assuming a life of five years and a resale value of $200,000,

a. Calculate the depreciation and the book value of this equipment for the next 7 years. Use both straight-line and double declining balance depreciation methods.

b. If the equipment is sold at year 4 for $400,000, what is the capital gain or loss using both of the above mentioned methods of depreciation?

4-An investor is purchasing a car repair facility with an initial cost of $90M. The facility will have before tax profit of $25M/Yr the first two years and $30M/Yr in years 3,4, and 5. At the end of year 5, it has to be disposed of at a cost of $10M with no resale value. The investor is using a straight-line depreciation method with an assumed life of 10 years and a resale value of $5M for depreciation purposes. He has a combined federal and state tax rate of 45%. His cost of money is 10%.

a- What is the net present worth of this operation?

b- Would he be better off using a DDB method of depreciation with the same assumptions?

5- An investor has two investment opportunities. Project A is a soft drink bottling operation (MACRS, three year class) and project B is a truck rental business (MACRS, five year class). The estimate of the financial situation of each project is presented in the following table. He also has the choice of investing the capital and receiving an interest rate of 15% annually. His total federal and local tax rate is 40%. In which project should he invest and why?

Data	Proposal A	Proposal B
Initial Cost	$50,000	$140,000
Salvage Value	$20,000	$30,000
Annual Benefit	$30,000	$40,000
Annual Cost	$15,000	$10,000
Life	3 years	6 years

6- A developer is given two options for acquiring a certain piece of construction equipment. His tax rate is 35%, his MARR=10%, and he uses a straight-line depreciation method. Which option should he take?

a.   Purchase at $100,000 and sell at 20% of its original cost after 5 years.

b.   Lease for $30,000 per year with no down payment.

7- A manufacturing company spends $400,000 for the purchase and installation of a fully automated production facility for a subsystem of its main product. This operation is going to save the company a net $100,000 each year. The company is using the straight-line depreciation method and assumes a life of six years and a resale value of $10,000 for depreciation purposes. The company abandons this operation after five years and sells the facility for $100,000. The company has a combined tax rate of 40% and his cost of money is 8%. Calculate the after tax NPW of this endeavor for the company.

8- A company purchased equipment at a cost of $120,000. Net income is estimated at $30,000/Yr. The estimated life of the equipment is 10 years, and it is estimated that the salvage value at the end of its life is $20,000. The company is using the straight-line depreciation method. The tax rate for the company is 25% and the cost of capital is 12%.

a.   Calculate and draw the after tax cash flow of this purchase for the next five years.

b.   If the company (which is profitable overall) sells this equipment at the end of the fifth

        year at a price of $35,000, and capital gain/loss has tax effects just as a regular income (i.e., at 25%), what would be the expected NPW of this project referred to at the time of the purchase?

9-A venture group is contemplating investment in either of the following projects:

a. Investment in cosmetic store with an initial cost of $100,000 and an annual net income of $20,000 The business is estimated to have a resale value of $300,000 after a four-year life.

b. Take over a beauty parlor with an $80,000 initial payment and an annual net income of $25,000 for four years. The lease will end at the end of the four years with no obligation on either side.

The group will pay you $2,000 to make a recommendation based on sound economic analysis. The cost of money is 20%, and the tax rate is 20%. Assume that the group is allowed to depreciate the initial cost in both cases and use the straight-line depreciation method with no resale value. What is your recommendation and why?

a. Use the NPW method and suggest which project the group should accept

b. If the salvage value for both systems is zero, use the Rate of Return method to make your suggestion.

c. If you only had paper and a pen, what do you recommend? Why?

10- In 1991 a newly created division of Fantastic Mechanical Works Inc. (FMW) purchased a machine with a total initial cost of $180,000 and put it in service immediately. The estimated revenue from this machine for the first three years is $58,000 with a $22,000 associated cost.

For the next three years the net income from this machine will be $ 36,000. FMW has an

effective income tax rate of 38% and uses a five-year life MACRS depreciation method. At

the end of the sixth year, the machinery can be sold for $30,000. FMW is a profitable

company and can apply tax losses from one division to tax payments of other divisions. FMW

has MARR=10%.

Calculate and plot the cash flow diagram for this project and calculate the before- and after-

tax NPW. If you have a computer or a financial calculator, calculate the before- and after-tax

rate of return as well.

CHAPTER 9

General Comments on Financial Analysis and Problem Solving

APPLICABILITY OF DIFFERENT METHODS

Any of the methods presented in this book can be used for comparing investment opportunities. The NPW is equal to the life cycle worth and is therefore the preferred method, especially when comparing the lifetime worth of systems or projects with equal lives. If acquisition of systems performing different functions is considered solely for financial gain, then the rate of return is the preferred method because it readily shows the relationship of the value of the project to the cost of money or the MARR. It can also be readily compared to the prevalent measure of benefiting from an investment, namely interest rate, or the safest available investment, which is government bonds. It is also a valuable measure for organizations borrowing money for investment. The disadvantage of this method is that the total cost, total benefit, and profit are not readily apparent from the analysis.

On some occasions the acquisition of one of several systems, all performing basically the same function, is definitely necessary regardless of the suitability of the investment in another type of business. In these cases the net present worth is the method preferred, especially in military, government, or social systems where almost equal benefits are obtained from different systems under consideration; therefore, the benefit is a non-differential factor. When comparison of

101

alternatives with different lives or different initial investments is considered, the uniform equivalent annual worth is more appropriate. An advantage of the EUAW is that we need not perform incremental analysis or equalize lives.

Benefit-cost ratio analysis is almost the same as net present worth. The advantage is that the ratio gives a perceived measure of percentage profitability. This percentage is sometimes mistaken with the rate of return.

Net future worth can be used when you need to know an operation or project will provide you with the money you need at a future date. This is especially helpful if you are saving for a future investment such as sending your children to college or for retirement purposes.

In using the different methods, care should be taken to perform incremental analysis or equalize lives when necessary. The following matrix is useful in making this decision.

Method	Incremental Analysis	Equalizing lives
NPW	NO	YES
C/B	YES	YES
EUAW	NO	NO
ROR	YES	NO

The above discussions on financial analysis and comparison of the alternatives using the cash flow model of Fig. 1.2 basically apply to deterministic engineering models, where the costs, the benefits, and their times of occurrences are known with a good degree of accuracy. In stochastic models, expected values of times, costs, and benefits are used.

In the derivation of the equations used in the financial analysis, the number of periods, n, is a whole number. In construction of the interest rate tables, n is also a whole number.

Therefore, for convenience of calculation, an end-of-period assumption is made, i.e., all the costs and benefits incurred during the period are assumed to occur at the end of that period.

ANALYZING FINANCIAL STRUCTURE OF AN INVESTMENT

The important part of the financial analysis is the construction of the cash flow diagram. The rest of the analysis consists of a set of calculations using mathematical formulae. The annual values of the cash flow diagram are taken directly from the statement of the problem and by applying the effects of depreciation, capital gain or loss, etc. After the cash flow diagram is constructed, the annual values can be entered into a spreadsheet, and by using proper commands, the computer will perform the required calculations. There are financial analysis calculators that can perform the calculations and provide the NPW, EUAW, and ROR when given the annual cash flow values. Students familiar with computers can construct their own special spreadsheet or program using any of the programming languages.

GENERAL STRUCTURE OF PROBLEMS

Every finite project, even if it involves replacement, has basically three stages:

   *. The beginning, that is the starting part.

   *. The operating period, which is the middle part.

   *. The close up, which is the end part.

Fig. 9.1 is the graphic presentation of the problem structure and illustrates the above parts.

There are important elements of a project that you have to understand and employ in financial

analysis in each of the stages mentioned above. These elements or factors are the basic

ingredients of a cash flow diagram and are shown under each of the stages in Fig. 9.1. Not all

of these elements are involved in all of the projects. For example, if we are considering a new

project, then there are no capital gains or losses to be considered in the beginning of the period.

On the other hand, in a replacement project where the opportunity cost has to be calculated, the

capital gain or loss at the starting point is an important factor. The list however is a reminder

of what factors may be involved.

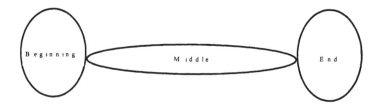

Beginning	Middle	End
Initial Investment	Operational Costs	Resale
Opportunity Cost	Operational Profit	Removal
Capital Gain or Loss	Depreciation	Capital Gain or Loss
Depreciation Method	Tax	Tax
Tax		

Fig. 9.1

The cash flow diagram is constructed using the information available and the above factors.

We then have to decide what measure to use in gauging the viability of the project, or in comparing two or more projects. If it is the NPW, then we have to bring everything to the front. If it is the NFW, we have to move everything to the end. If it is the EUAW, we have to spread everything in the time span of the project. Finally if our measure is the ROR, we have to assume the interest rate as an unknown that makes NPW = 0.

EXTRA PROBLEMS

The following is a list of problems for giving the user of this book an extra opportunity to

sharpen the application of financial analysis tools.

1-Production equipment is bought at an initial price of $10,000. The annual operation and

maintenance cost is $100. The salvage value at the end of the 15-year life is $500.

    Using MARR of 10% calculate:

        a. the net present worth

        b. the equivalent uniform annual worth

2- Equipment is bought for an initial cost of $20,000. Its operation will result in a net income

of $6,000/Yr for the first year, increasing by $1,000 each year after year one. At the end of

the fifth year the equipment is sold for $5,000. The prevailing interest rate for the next five

years is estimated at 10%

a. Draw the cash flow diagram for this project.

b. For this project calculate

        1.   The NPW

        2.   The EUAW

        3.   The NFW

c. Another model of the equipment with the same initial price and annual cost brings in an income of $1,100 per month but has no salvage value. What is the rate of return for this project?

3- The following two projects are recommended for investment:

     a. Initial investment $800K, annual income $400K

     b. Initial investment $1,500K, annual income $700K

The lives of both projects are five years. The projects are in a tax-free zone and the inflation is considered to be negligible. The investor has a MARR of 30%. He comes to you for advice and asks you to use the ROR method and make a recommendation. What would you recommend?

4- Mr. "X", a friend of yours, is asked to invest in the following project: installation and operation of a facility with a life span of five years. The initial investment is $90M. It will have a net profit of $25M/Yr the first two years and $30M/Yr in years 3, 4, and 5. At the end of year 5 it has to be disposed of at a cost of $10M with no resale value.

a. If he has the money and his opportunity cost of money is 10% (i=10%), would you advise him to invest or not? Yes? No? Why? Explain.

b. If he can only put down $33M but his bank will extend him a loan for the rest of the initial

investment to be used on this project, at 15% in a way that he has to pay it back in equal

installments at the end of each of the five years with no collateral, would you advise him to

take the loan and invest? Yes? No? Why? Explain.

5-An investment group is considering building a multilevel parking garage in the downtown of

a city. They received three proposals. Proposal A is for building a 2-level garage, proposal B

is for a 4-level garage, and proposal C is for a 6-level garage. The financial situations

estimated for the three proposals are shown below. Their cost of money is 10% annually.

Using the NPW method, perform the financial analysis and make your recommendation as to

which of the proposals, if any, they should accept.

Data	Proposal A	Proposal B	Proposal C
Initial Cost	$60,000	$100,000	$150,000
Annual Revenue	$52,000	$75,000	$80,000
Annual Expense	$22,000	$25,000	$27,000
Intended Life	3 years	6 years	6 years
Estimated Resale	$30,000	$40,000	$45,000

6- A Midwestern state is considering building a stretch of 2-lane toll highway between two

points. It will cost $2 million and will bring an annual toll revenue of $300,000 a year. The

cost of operation and maintenance will be $6,000 annually. If they build a 4-lane divided highway, the cost will be $3 million but will increase the toll revenue to $410,000 a year. The cost of maintenance and operation of the 4-lane highway will be $10,500 a year. They use a 15 year analysis period and 9% MARR. Should they build a 4-lane or a 2-lane highway? Perform the financial analysis using the rate of return method.

7- An investor in Maryland is offered two investment opportunities. Project A is to invest in a small printing operation consisting of a high-speed printing press. The initial cost of the press including installation is $60,000. The press costs $18,000 per year to operate and will bring in an annual income of $48,000. However, the productive life of this press is only three years and can be sold at $24,000 at the end of year 3 to a secondhand dealer. Project B is a small taxicab operation involving four cars to be purchased at $30,000 each. The total operation cost of the four cars is $40,000 per year and the total annual income is $72,000. The useful life of the taxicabs is six years each and their salvage value is $9,000 each. If the MARR of this investor is 10%, which one of the projects if any should he accept? Use the NPW method.

8-Thrift Transportation Company wants to bid for a new business involving transporting passengers in a new line between two cities. The company is studying the size of the bus needed for purchase for this new route. The company has three choices:

a. Purchase a 20-passenger bus with a price of $49,500 that has an operating cost of $40,000 per year, an estimated life of three years, and a salvage value of $10,000.

b. Purchase a 40-passenger bus with a price of $107,000 that has an operating cost of $75,000 per year, an estimated life of five years, and a salvage value equal to the cost of its removal.

c. Purchase a 50-passenger bus with a price of $151,000 that has an operating cost of $95,000 per year, an estimated life of six years, and a salvage value equal to the cost of its removal.

Assume that the bus in all the cases will operate with a full load and that the passenger seat annual income is $3,000 (i.e., each passenger capacity will bring an income of $3,000 per year). If the transportation company has MARR of 15%, which of the buses if any should the company buy? _Use the ROR method._

9- An investor is being asked to invest in a project with an initial investment of $3,000 with first year income of $400 increasing by $100 every year for five years. His MARR is 5%. If he phones you with this problem when you have no access to anything other than paper and pen, what would be your recommendation to him? Should he accept this proposal or not? Explain your answer.

10-An investor is offered the opportunity of investing $100,000 in a six year project that will have a net income of $25,000 per year.

a. If inflation is a constant 4% per year and his after inflation MARR is 10%, should he invest in this project?

b. If he is looking to invest his $100,000 where he can get an inflation adjusted (i.e., net, after inflation is taken into account) of 12%, what minimum rate of interest he should shop for?

11- Hosbol Corporation purchased a system for $1 million. The net income from operating this system is $300,000 per year. Assuming a life of five years and no salvage value, what is the Net Present Worth (NPW) of this system? Hosbol uses the double declining balance (DDB) depreciation method, its cost of money is 12%, and its tax rate is 33.33%.

12-The initial cost of equipment is $1,000,000. Assuming a life of five years and a resale value of $200,000,

a. Calculate the depreciation and the book value of this equipment for the next seven years. Use both straight-line and DDB depreciation methods.

b. If the equipment is sold at year 4 for $400,000, what is the capital gain or loss using both of the above mentioned methods of depreciation.

13- An Idaho farmer buys farm equipment for $120,000. He is using DDB depreciation. The intended life of the equipment is four years and the estimated resale value is $20,000. If he sells the equipment at the end of the third year for $15,000, what is the capital gain or loss?

14-Tacoma Shipbuilding Company (TSC) in Tacoma, Washington, is purchasing a machine at $165,000 for a plate-forming project. Transportation and installation will add $15,000 to the initial cost. The total initial cost is to be depreciated using a five year life DDB method with an assumed resale or salvage value of $40,000. The yearly income from this project for the first year is $60,000 with an associated expense of $35,000. In the second year, the income increases to $70,000 with the same expense as the first year. In year 3 income is $80,000, and increases by $5,000 each year thereafter. Associated expenses remain at $40,000 after year two. The operation is terminated at the end of year 5, and the equipment is sold for $35,000. If TSC has a tax rate of 35% and MARR of 12%, what is the NPW of this operation? Assume as we always do, that the TSC has other profitable operations.

15- ABC manufacturing has a project involving the purchase of equipment for $35,000 that will increase sales of ABC by $14,500 per year. The annual operation costs of this equipment for the first three years are $1,000, $2,000, $4,000, respectively. The costs will increase by $3,500 every year from then on. The equipment is sold for $5,000 at the end of year 5. The MARR for the owner of this equipment is 10%, their tax rate is 20%, and they use straight-line depreciation assuming a four-year life and a $10,000 resale value. Determine the after-tax cash flow of this project. What is the after-tax NPW of this project? ABC has other projects going and is very profitable.

PART 2

LIFETIME WORTH (LTW) ESTIMATION AND CALCULATION

CHAPTER 10

Lifetime Worth: Background and Definitions

BACKGROUND

Until a few decades ago, purchase price was the only consideration when a decision was made in regard to the purchase of a new item. The item could be as ordinary as an automobile or as sophisticated as a warship or an air traffic control system. As systems became more complex and as funds, public or private, became scarcer, costs, such as operational and maintenance costs, became important considerations. The natural extension of these considerations was the notion of system life. How long should we be concerned about the operation and maintenance costs? When do we want to replace or retire the system? What is the system's life span? What is the intended life, i.e., how long do we plan to keep the equipment or system in service? And, what is the estimated technical life, i.e., how long will the system stay operational before its technical characteristics fall below an acceptable level? Consideration of these questions gave rise to the subject of Lifetime Cost (LTC). The importance of LTC on the customer side forced its consideration on the producer side. Traditionally, the producers only considered initial price and quality as items of competitiveness- now LTC became an important competitive item to consider. Consideration of the lifetime cost originated in the military and defense related systems. It reflects the notion that military items have only costs associated with them and not any profits. This was in the days that benefits were a non-differential factor, were not of concern, or could not easily be converted to monetary terms.

115

The lifetime benefits of a system are measurable and important. Even the benefits of military systems can and are being measured in dollars. System productivity is an important factor in decision making involving alternatives or in the treatment of replacement decision. Even if you are comparing different missiles or considering replacement of a system by a new system, the productivity is a benefit that should come into the analysis procedure. It is therefore appropriate and timely to use lifetime worth (LTW) in place of lifetime cost.

The term "lifetime worth" reflects the net gain or loss associated with the total life of the system. In financial analysis, we are mostly concerned with the worth rather than the cost. Therefore, we will generally use LTW except where costs are the only items to be considered. In this particular case, we can use LTW or LTC interchangeably.

Lifetime worth is used by the system buyers to decide on alternative courses of action in acquisition, by system owners when analyzing replacement problems, and by manufacturers when making alternative design choices.

Table 10.1 presents the acquisition cost of several items versus the costs of their operation and maintenance. It also states the average length of time these costs are calculated for.

Table 10.1

Product	Time Period (Years)	Capital Investment	Opr. & Maint.
Elect. Washer/Dryer	12	73	27
Electric Range	15	38	62
Full-Size Car	4	70	30
House	30	70	30

A General Accounting Office (GAO) study has shown that the operating cost of a hospital in the first three to five years exceeds the cost of its construction.

LIFE

The length of time we consider in calculating the lifetime worth depends on the point of view and the purpose of the calculation. In the following sections, we define duration of life from several points of view. Either of these definitions can be used for different purposes. For our purpose, we use the word "system" for any entity that performs a function. It has input, applies a process to it, and produces an output. For the purpose of the lifetime worth discussion, we will use the word system for any system, equipment, program, or project.

SYSTEM LIFE

We define this as the time span from initial conception of the idea of producing a system, to the time when for whatever reason the system no longer serves the purpose for which it was intended. It is then retired or replaced.

It is important to distinguish the difference between the function and the purpose. A system can still be functional, yet it might have served its useful, intended purpose and should be removed from service. A helicopter is required to provide access to a mountaintop observation site. After the completion of the site, an access road is built; although still functional, the helicopter has served its purpose. It is no longer required and has to be disposed of.

PHYSICAL LIFE

Physical life is defined as the duration from the time a system is up and starts producing output to the time its characteristics fall below the acceptable level and cannot function as intended. This assumes that it technically cannot be brought back to the acceptable functional level.

TECHNOLOGICAL LIFE

This is the span of life after which the system becomes technologically obsolete for the purpose for which it is being used. We have seen this happen to personal computers in recent years. For Intel 8086 based systems, technological life was about three years. This was reduced to about one year for 8386 and became much shorter for 8486, Pentium, and what followed.

ECONOMIC LIFE

The economic life starts at the time an investment is made to acquire the system and continues until the time when it is no longer profitable to keep the system in service. The economic life depends on the cost benefit "time profile" of the system. As we will see later, it ends when the equivalent uniform annual worth (EUAW) is at a maximum or a minimum. The main difference between the physical life and economic life is that different points in time for the start and the end of the system life are determined. Therefore, total life costs and benefits are affected. In general, the system's physical life starts after the economic life, since an investment is usually made before the system is put into service. The ends of the economic and physical lives may or may not coincide.

LIFE OF INTEREST AND PLANNING HORIZON

The financial analysis of any ongoing concern over a period of time requires detailed knowledge of all the costs and benefits associated with its operation and throughout its "life of interest". The word "interest" is injected here on purpose. We may be interested in analyzing the costs and the benefits of a certain portion of the life of the system. This may be all or part of the total life of the system; hence, this definition of life is different from definitions mentioned earlier. To the decision maker in the Department of Defense (DOD) in charge of a new cargo aircraft acquisition, the life of interest begins at the time he allocates the cost of developing the concept and continues to the estimated time of retirement of such aircraft from service (perhaps the end of its economic life). To the organization purchasing the same aircraft from the Department of Defense for commercial use, the life of interest starts at the time of purchase. Therefore, the life of interest depends on how long we are planning to use the system, in other words, the planning horizon. However, lifetime worth is normally considered to be the NPW of the total costs and benefits associated with a system from the time the initial capital investment is made to the time it is removed from service.

PHASES OF THE LIFETIME

The breakdown and analysis of the lifetime components are important in the study of the changes in the costs and benefits of the system throughout its lifetime. The life span of a system can be divided into several different phases. The division of the system life into phases is mainly done for the purpose of separating the costs and the benefits that are associated with the phases. The beginning and end of these phases are not always well defined, and the phases sometimes slightly overlap. Nevertheless, an attempt to separate and define these phases will help the analysis and modeling of the lifetime.

The major phases of a typical system life with their cost profile are shown in Fig. 10.1. The relative ratios of the costs in Fig. 10.1 are gross generalizations. They are estimated from the experience of the author and the values given in the published literature. It is quite obvious that the actual values depend on the particular system under consideration. The non-recurring costs and the total units produced make the ratios case dependent. Some other costs, such as the cost of project management, marketing, and warranty, etc., are spread over the total life of the system. These are called general and administration (G&A) costs and overhead. They are embedded in the cost of each phase as an added percentage.

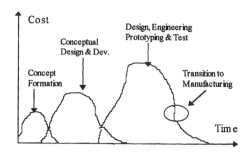

Fig. 10.1

PHASE 1- CONCEPT FORMATION

In this phase, the idea of developing or acquiring a system is formed and the basic concepts are formulated. The idea is generated from a need analysis or some other source such as marketing, customer feedback, business intelligence, etc. Normally the cost of this phase is minimal compared to the total cost unless it would involve purchase of specific intellectual property.

PHASE 2- CONCEPTUAL DESIGN AND DEVELOPMENT

In this phase alternative design philosophies are evaluated and the use of different technologies is considered. Maintenance goals such as mean time between failure (MTBF), mean time to repair (MTTR), availability, reliability, and maintainability are set. The conceptual design produced in this phase defines the final form of the system.

PHASE 3- DESIGN, ENGINEERING, PROTOTYPING, AND TEST

The final configuration of the total system is completed in this phase. Usually a system prototype is produced and tested. Modification as the result of the test outcome is incorporated into the system. Production, operation, and maintenance documentation are produced, and the major part of system management is transferred to the manufacturing division.

PHASE 4- MANUFACTURING, TEST, PACKAGING, AND DELIVERY

In this operation the system is manufactured, tested, prepared, and packaged in a form ready for final delivery to the customer. The activities involved in this phase and their costs depend on whether we are considering a unique system or production of several identical units.

Examples of unique systems include air traffic control, large communication systems, air defense systems, automobile factories, etc. We call these made-to-order systems. The costs of manufacturing for these systems include the costs of purchasing the individual components of the system, integrating, testing, and often, installing. The total cost of all of the phases up to this point is the initial cost of the system. To this cost, the producer adds a profit and calls it the system price. In our financial analysis language, from the customer's point of view this is the initial investment.

Some items such as automobiles, television sets, and aircraft (military or commercial) are produced in quantities. From the buyer's point of view, these are ready-made items. The quantity can be in one digit number, tens, thousands, or millions. For these items, there are two basic costs. Fixed costs (FC) and variable costs (VC). Fixed costs are those costs that are not dependent on the number of production units. Costs of phases 1,2, and 3 are of this type. Variable costs are dependent on the number of units manufactured. Material and labor are examples of this type. The total cost of a production unit is, therefore, the total of the fixed and variable cost divided by the number produced. The cost to the customer at this point is the unit's cost plus any profits added to the cost. This is his initial investment.

$$UnitCost = \frac{FC + VC}{N} = \frac{FC + aN}{N} \qquad (10.1)$$

where a is the variable cost associated with one unit, and N is the number of units produced. Fig. 10.2 presents the relationship of fixed cost, variable cost, unit cost, and the number of units produced. As expected the total cost of one unit decreases when the number of units increases. However, we have to be careful not to apply this principle indiscriminately. When the number of production units goes beyond the capacity of the system, then additional fixed costs occur in increasing the system capacity to accommodate the additional units. In this case, we have to go back and calculate the relevant cost associated with the new production capacity.

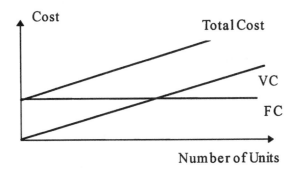

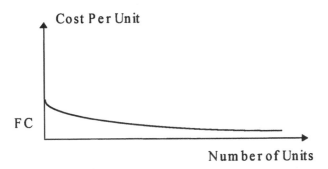

Fig. 10.2

There are also other costs associated with purchasing activity including requirement generation, bid document preparation, evaluation of bids, and other necessary activities. These costs can be included in the initial investment.

PHASE 5- INSTALLATION, OPERATION, AND MAINTENANCE

The operation and maintenance cost of the system constitutes a major portion of the lifetime cost. During this phase, the system starts generating benefits. Some attributes of the system such as MTBF, MTTR, reliability, maintainability, etc., have considerable influence on driving the cost of this phase.

PHASE 6- REMOVAL, SALVAGE, OR RESALE

When, for whatever reason (retirement or replacement) the system has to go out of service, a cost is associated with its removal. If it can still be used by another organization or converted to another form, it may have a resale or some salvage value that may compensate or even be higher than the removal cost. This expense or income will have an effect on the comparison of the lifetime worth of systems.

CONCURRENT ENGINEERING

As mentioned previously, the phases of the lifetime mentioned above are not quite separated from each other. There are overlaps, and there are time periods when the costs for more than one phase have to be taken into account (Fig. 10.1). This fact has to be considered when constructing the cash flow diagrams in the next section. Concurrent engineering, that is, developing some of the phases concurrently, may reduce the total system cost and the total development time.

PROBLEMS

1- Investigate the lifetime stages of a residential property and attach a cost to each stage as a percentage of the total cost.

2- You are the owner of a high-rise office building that for prestige reasons is very much in demand. How would you construct your rent? Make any assumption you want.

3- A newly incorporated company intends to produce plastic containers. A machine with a production rate of 200 containers per hour has an initial cost of $50,000. Other costs associated with the production are

> Raw material for one container--$5.
>
> Machine operator --$10 per hour for 8 hours a day production
>
> Hourly electricity used by the machine--$1/hour
>
> Building to house the offices and production facility--$150,000
>
> Personnel salary, advertisement, utility, and other costs

If the containers can be sold at a price of $10 per unit, how many containers should the company sell to break even? Do not consider depreciation and tax.

4- A car wash at the corner of Willshire and Olympic boulevard in Los Angeles is for sale for $500,000. Mr. Buycheep who intends to buy the facility estimates that it will have a 100 car per day customer base. The cost of water, electricity, washing liquids, etc. per car is $1. For administration and operation of the facility, three people at $7 per hour must be employed. If the purchaser wants to take home $300 per day and recover his investment and costs in two years, what should he charge per car wash?

CHAPTER 11

Lifetime Estimation and Calculation

THE CASH FLOW DIAGRAM

The cash flow diagram is the pictorial presentation of inflow and outflow of money as the result of the operation of the system. By convention, the outflow is represented by downward arrows and the inflow by upward arrows.

The first cost associated with the system is the procurement cost. In a made-to-order system (e.g., military systems), the procurement cost is composed of the cost of phases 1,2, and 3 of the lifetime that occur at different times. The cost of procurement in a ready-made system is the purchase cost and any other administrative cost associated with the purchase and occurs at the time of purchase. After a system is put in service, there are various costs associated with its operation and maintenance. The system will also start producing some benefits. In some systems such as government projects, the benefits may not easily be measurable in monetary terms. For comparison purposes, a subjective monetary value can be attached to these benefits. The system may also have a salvage value when it is taken out of service. Depreciation and tax also affect the inflow and outflow of cash. The time profile of these inflows and outflows (the cash flow diagram) is a useful and essential tool for the financial analysis required in the decision making.

Fig. 11.1 exhibits the various costs and benefits associated with the different phases of the system lifetime for a ready-made system.

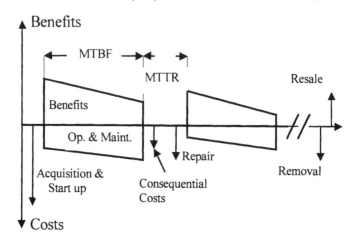

Fig. 11.1

It is important to ensure that all tasks or activities involving costs and benefits are considered, accounted for, and included in the cash flow diagram. The following points should be considered when analyzing Fig. 11.1:

(1)   The acquisition cost includes the purchase price plus the cost of all other activities that make the system operational. For a made-to-order system, this is the present value of all the costs incurred for the life phases up to this point.

(2)   Operation and maintenance includes preventive maintenance (PM) but not major overhaul or repair during failure.

(3)   Repair costs include all costs of overhaul and major repair including all costs of lost production, e.g., idle workforce.

(4)   It is assumed that the system is down and nonproductive during the repair period.

(5)   Consequential costs are the costs of associated systems taken out of service during the repair time.

(6)   Reduction in productivity and increase in operational costs as the system ages are to be obtained from regression models based on statistical data.

(7)   The values shown are after-tax values.

The model presented in Fig. 11.1 can be used for calculating the system lifetime worth. It can also be used to compare the merits of different systems. If all the costs and benefits are accounted for, then financial methods discussed in Part 1 can be used for this purpose.

WORK AND COST BREAKDOWN STRUCTURE (WBS-CBS)

Top level work breakdown structure (WBS) exhibiting the activities in each phase of the system life is shown in Fig. 11.2. This structure is obtained by first specifying what activities have to be performed to complete the job or the project and then connecting them in a logical manner starting from the endpoint, that is, from what we have to have at the end. We then have to obtain the detail resources required to complete each activity and the corresponding total cost of resources in monetary terms. When we attach the total cost to each activity, we will obtain the cost breakdown structure (CBS) which is the cost counterpart of WBS. When the project starts to perform there, it usually provides some benefit. The complete time line of the costs and benefits will provide the cash flow diagram from which the total lifetime worth of the project can be calculated. The combined work breakdown structure and the cost benefit structure will help identify all the costs and benefits associated with the development, production, and operation of the system. CBS is often used by the manufacturers to estimate the unit cost or the lifetime cost of the system.

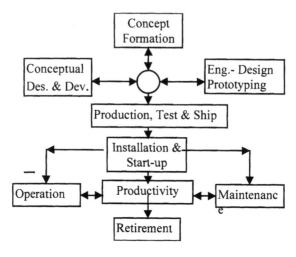

Fig. 11.2

Table 11.1 shows the levels of CBS. It shows the related items associated with each phase of the life in more detail. Each cost is related to an activity. Concurrent engineering allows the performance of the activities in parallel rather than sequentially.

LIFETIME WORTH CALCULATION

If we have the detail of all the costs mentioned in Table 11.1 with their time of occurrence, we can construct the system lifetime cash flow diagram (Fig. 11.1) and calculate the NPW of the system using an appropriate MARR. We have to remember that we are estimating these costs and benefits and their rate of change with time. Therefore, the lifetime worth is also an estimation. However, in calculating the LTW in this manner, we have included the effect of time value of money. Another advantage of calculating the LTW using the cash flow is that the effects of depreciation and tax can also be taken into consideration. As we see later in other methods of estimation, time value of money, depreciation, and tax effects do not enter into calculation. The reason is that most of the organizations using these methods of estimation, mostly military, are tax exempt and do not borrow money for their initial investment. With the budget deficit and the national debt being what they are today with no relief in sight, this is going to change. Already government organizations are required to perform cost-effective investigations of their projects. They are also asked to take the interest rate into account in evaluating alternatives. Soon other estimating methods have to be developed that will take tax, depreciation, and interest rate into account when estimating LTW.

Table 11.1 - Generic Work Breakdown Structure

* **Concept Development:**
    * Needs analysis
    * Requirement development
    * Develop operation and maintenance concepts
    * Develop general system specifications and block diagram
* **Conceptual Design and Development:**
    * Generation of operating specifications
    * Generation of physical specifications
    * Proof of the applicable new technologies
    * Development of the general system design (GSD)
* **Engineering, Design, and Prototyping:**
    * Detail system design and validation
    * Subsystem/unit design and engineering
    * Preparation of drawings and material list
    * Prototype fabrication and test
    * Final working drawings and material purchase list

* Production:

    * Preparation of production plan

    * Material purchase and storage

    * Purchase of capital equipment specific to the system

    * Kiting of parts

    * Fabrication and assembly

    * Burn-in and test

    * Packaging and shipping

* Installation and Start-up:

    * Site preparation

    * Building

    * Test and acceptance

    * Initial training

* Operation:

    * Energy

    * Supplies

    * Personnel (labor)

    * Replacement training

* Maintenance and Repair:

    * Training and replacement training

    * Equipment

    * Inventory (spares, etc.)

    * Lost production

* Retirement:

    * Deactivation and shutdown

    * Removal (salvage)

* Management (all phases):

    * General management (G&A, warranty, cost of money, etc.)

    * Program management

Example 11.1

XYZ Corporation purchases fully automated (no labor) equipment needed for its production line. The cost of purchase, installation, and start-up adds up to $100,000. The MTBF and MTTR are stated by the manufacturer to be 11.5 months and 15 days, respectively. The cost of operation is $5,000 per month increasing by $250 per month per year. The administrative and nondirect expenses allocated to the operation of this machine are $5,000 per year increasing by $500 each year. No routine maintenance is required, and the cost of major repair and overhaul is $3,000 per year. The output produces a gross income of $10,000 per month reducing by $200 per month each year due to loss of productivity caused by aging. For a planned life of six years, calculate the LTW of this machine. The company uses straight-line five-year life depreciation, its total tax rate (federal, state, and local) is 40%, and its cost of money is 12%. The equipment has to be scrapped at the end of the sixth year. Its salvage value is equal to the cost of its removal. What is the LTW of this equipment?

Solution:

Depreciation: $\dfrac{100,000}{5} = 20,000$ per year

First year: Expense:

Direct Operation	$5000 * 11.5 = 57,500$
Major Maintenance	3,000
Administrative	5,000
Total	65,000
Income	$10,000 * 11.5 = 115,000$

Years 2 to 6

Increase in expenses per year $250 * 11.5 + 500 = 3375$

Decrease in income $200 * 11.5 = 2300$

We can, therefore, construct a table similar to the table of Example 9.5 and develop the after tax

cash flow and calculate the lifetime worth.

Yr.	Income	Expense	OP.	Depr.	Bk.	Taxable	Tax	IAT
0		100,000	-100000					-100,000
1	115,000	65,500	49,500	20,000	80000	29,500	11,800	37,700
2	112,700	68,875	43,825	20,000	60000	23,825	9,530	34,295
3	110,400	72,250	38,150	20,000	40000	18,150	7,260	30,890
4	108,100	75,625	32,475	20,000	20000	12,475	4,990	27,485
5	105,800	79,000	26,800	20,000	0	6,800	2,720	24,080
6	103,500	82,375	21,125			21,125	8,4500	12,675

Fig. 2.5 is the cash flow diagram representing the financial situation of this equipment through its

intended life.

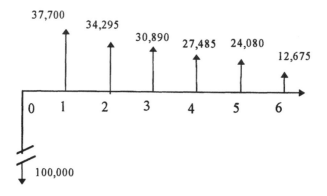

Fig. 11.3

$LTW = -100000 + 37700 (P/F, 12, 1) + 34295 (P/F, 12, 2) + 30890 (P/F, 12, 3) + 27485 (P/F, 12, 4)$

$+24080 (P/F, 12, 5) + 12675 (P/F, 12, 6) = 20,547$

Spreadsheet programs can easily be developed to calculate the LTW. The following spreadsheet

will provide us with the same result. The advantage is that we can do a sensitivity analysis using

different interest rates or tax rates. The sensitivity analysis allows us to understand the risk

involved if actual tax rate or cost of money is not exactly what we estimated.

	A	B	C	D	E	F	G	H	I	J
1	Example 11.1									
2										
3		Cost	100,000			Resale				
4		Years	5		Estimate	0	Iterest	12%		
5		Good wil	0		Actual	0	Tax Rate	40%		
6					Depreciation SL	2,000				
7										
8		Year	Income	Expense	OP	Depr.	Bk. Value	IBT	Tax	IAT
9		0	0	100,000	-100,000					
10		1	115,000	65,500	49,500	20,000	80,000	29,500	11800	37,700
11		2	112,700	68,875	43,825	20,000	60,000	23,825	9530	34,295
12		3	110,400	72,250	38,150	20,000	40,000	18,150	7260	30,890
13		4	108,100	75,625	32,475	20,000	20,000	12,475	4990	27,485
14		5	105,800	79,000	26,800	20,000	0	6,800	2720	24,080
15		6	103,500	82,375	21,125			21,125	8450	12,675
16										
17		Year	0	1	2	3	4	5	6	
18		Cash Fl.	-100,000	37,700	34,295	30,890	27,485	24,080	12,675	
19		* Net cash flow in the last year is the sum of the after tax vlaues of the income from								
20		The operations and that of the good will.								
21		* Now you can use the NPW and IRR spreadsheet formulae as in chapters 2 to 5								
22						NPW=	$20,457			

LIFETIME COST ESTIMATION

More often than not the amount and time of occurrence of the costs and benefits

associated with the system life are not known before the system is produced. Therefore, the

financial analysis methods for calculating lifetime cost using cash flow cannot be used. This

presents a problem, and we have to somehow estimate the lifetime cost. The estimations are

normally based on historical data gathered from the same or similar types of systems or from the estimates of individual tasks and activities that have to be performed for each element of the work breakdown structure. The data gathered and the analysis and classification required to produce the estimated costs and benefits are very important if a reasonably accurate and reliable estimation of LTW is to be made. Three methods of estimation are normally used to produce the lifetime cost and to estimate the benefits obtained when the system is put into operation.

PARAMETRIC ESTIMATION

In this method, the historical data are obtained, classified, and tabulated as a function of one or more attributes or parameters of the system. The categorization and classification are very important in order for them to be universally applicable to LTC estimation. Parameters such as system output, weight, size, speed, accuracy, complexity, etc., are chosen as the independent variables. The cost of each element of the lifetime has to be related to one or more of the parameters mentioned above or any other appropriate parameter. Regression analysis is used to construct the appropriate relationship between the LTC and these parameters. These are called Cost Estimating Relationships or CERs. Historical data for this purpose are available for some specific systems types such as a building where construction and maintenance costs per square foot for different types in different areas are known with accuracy. We could construct a one parameter estimation, e.g., attach the cost to the square footage. Statistics collected by a constructor in Northern Virginia (Mr. Golegolab) conclude that low-cost residential building costs are $52 per square foot. Maintenance costs are $0.5 per square foot per year (present value). The average life is 30 years. Hence, a 1,600-square-foot unit's lifetime cost is

$$1600 \times 52 + 1600 \times 0.5 \times 30 = 107,200$$

If more than one parameter is used, a weight has to be attached to each parameter. We can go into further detail and use extended data to estimate the cost for each type of residential building and each state or county. If we want to use multiparameters, we can subdivide the building into main structure, electrical wiring, air conditioning, etc. Estimate the value of a unit of these and give a corresponding weight to each of them according to their percentage of the total cost of the building. Insurance companies often use publications such as Boeckh publications to estimate the replacement costs of real properties they insure.

The following table presents the cost of each subsystem of a residential building as a percentage of its total cost. If the cost of a subsystem can be estimated with reasonable accuracy, the total cost can be calculated using the related percentage. In the case of a residential building early in the design phase, the architect can determine from experience or in some cases with good engineering accuracy the cost of one of the subsystems. He/she can then estimate the total cost of the building. For example, if he is designing a moderate residential house with three bedrooms, the cost of an air conditioning system is about $10,000. He can then use Table 11.2 to estimate the total cost as

$$\text{Total Cost} = 10000 \cdot 0.0649 = 154{,}0832$$

This is just the cost of the building exclusive of land, site preparation, etc.

Segment	% of the total cost
Main Structure	57.8
Plumbing	6.52
Electrical	5.45
Heating and Air Conditioning	6.49
Kitchen and Appliances	4.54
Site Preparation	2.36
Indirect Cost and Overhead	16.84

Table 11.2

For some other systems such as cargo vessels, cost elements are given as a function of tonnage and available historical data. As in the case of the building, a multi-parameter cost calculation can also be made. For computer software systems, the estimating parameter is the system size or source line of code (SLOC) that provides an estimate of the number of hours required to produce a system software. The productivity of the software engineers in number of codes per hour also enters the estimation. The maintenance cost of a software system also needs to be estimated to complete the lifetime worth. The lifetime of the software system depends on the type of the software. Software systems mostly become obsolete due to emergence of new software with better functions or new tools that make the software more user friendly. Because of the importance of the calculation of the software LTW, we will discuss this subject in greater length in a separate chapter. In some systems, especially if the acquisition cost is known, other costs as a percentage of acquisition cost are relatively easily obtained from historical data. In some others, we have to attach the estimates to parameters that are available early in the lifetime, such as the required output in the case of a simple generator or other design parameters such as resolution and probability of detection of a radar or the speed of calculation as in case of computers. These relationships can be obtained from historical data but change with the advancement of science.

COMPARATIVE ESTIMATION

Here we calculate the lifetime cost by comparing our system with a similar system that has already been in service for some time. We know with sufficient accuracy from historic data that the total lifetime cost and the cost of the different elements of a power station change as its MW capacity changes to the power of 2/3.

$$\text{Cost} = \text{K (MW)}^{2/3} \qquad\qquad (11.1)$$

Let us say that we know the lifetime cost of a 300 Mega Watts non-nuclear power station to be C_1. We now want to estimate the lifetime cost of a smaller power station, e.g., 100 MW.

$$C_1 = K (MW_1)^{2/3}$$

$$C_2 = K (MW_2)^{2/3}$$

$$C_2/C_1 = (MW_2/MW_1)^{2/3}$$

$$C_2 = C_1 (1/3)^{2/3} = 0.48 \, C_1$$

A more accurate estimation is to obtain empirically the functional relationship of the change in cost with the change in power for each element of cost. After each cost element is estimated, the total lifetime cost is easy to estimate. The more accurate the relationship, the more detailed the elements of cost, and the better the estimate. The table below shows the cost of different types of power plants that can be used for cost estimation.

Type	Size	Initial Cost
Coal Fired	600 Mega Watts	$1600/KW
Gas Turbine (Gas or	150 Mega Watts	$500/KW
Nuclear	1000 Mega Watts	$3000/KW

ENGINEERING ESTIMATION

In this method, the work breakdown structure is used as the basic work sheet. The more detailed the WBS, the more accurate the estimate. After the WBS is constructed, engineers and estimators will break each task or event into labor, material, equipment, G&A, and other constituting elements. They will then estimate the cost of each task using the prevailing market cost for each element. After this is done the rest is simple, and the estimate of the total LTC is made by adding all the costs.

COMBINATION OF THE ESTIMATING METHODS

A combination of these methods can also be used depending on what type of system is under consideration, what part of the lifetime we are concerned with, and what kind of historical data we have available. We can start the process with a parametric or comparative estimate for budgetary purposes and move into an engineering estimate as the work progresses and more detail becomes known about the system. Later on into the process, we may have enough data to construct a cash flow diagram and do a more detailed calculation.

ESTIMATION ACCURACY

The accuracy of estimates ranges from 60% to 95%, depending on a lot of factors. Engineering estimates are the most accurate estimates, and parametric are the least accurate. Comparative estimates are in between the other two. The parametric and comparative methods rely mostly on historic and relational data. The rapid technological and other changes during the years affect the reliability and the applicability of the historical trends. This can be overcome by carefully watching the advancements in the related sciences and technologies, evaluating their influence in different phases of the system lifetime, and incorporating them in the process of cost estimation. The engineering estimate is more technical. If the estimators have good background and experience in the particular task, and accurate, up-to-date knowledge of the market prices, a reasonably accurate engineering estimate can be made.

UNIVERSAL RATIOS

There is always a temptation to produce some comparative ratios of the costs of different phases of the lifetime or to obtain the cost of each phase as a percentage of the total or acquisition cost. A reasonably credible set of historical relationships from which an estimate of these ratios or percentages for specific system categories, especially for the construction industry, is available. The technological changes mentioned previously have effects on these ratios that have to be considered. Unfortunately, all encompassing or universal ratios or percentages depend on numerous factors, and reliable numbers are not available. The cost of research and development is spread over the number of systems produced. Therefore, the percentages for an air traffic control radar are very different from those of a television set, although both can be considered electronic equipment. The maintenance costs of the same systems vary considerably depending on the closeness and efficiency of the maintenance facility. The cost of slow-moving spares depends on how many systems are maintained at one facility.

The design and trade-off considerations discussed previously and system sophistication have an effect on the distribution of the lifetime cost between the phases of the system life. A set of universal ratios or percentages would be misleading, but subjective relationships, such as change of per copy development cost with lot size or relationships between maintenance cost and sophistication can be helpful in lifetime cost estimations.

PROBLEMS

1-The initial cost of purchase and installment of an automated shoe-making machine is $100,000. The indirect cost of operation of the factory allocated to this machine is $2,000 per year. The machine operates 4,000 hours each year and produces one shoe every 0.5 hour. The maintenance cost per year is $1,000. The shoes are sold to distributors at $10 each with 10% net profit. If the machine lasts five years and has a removal cost of $1,000 with no salvage value, what is the lifetime worth of this machine? Use a 6% annual interest rate.

2-The purchase costs of cargo ships are estimated to be $800 per ton. A shipping company purchases a 30,000-ton cargo ship for transpacific operation. The ship carries 20,000 tons of cargo per month and stays at the dock one month a year for major repair that costs $20,000. The shipping company has operational net income (OP) of $20 per ton. The administrative cost allocated to this operation is $30,000 per month. At the end of a 20-year life, the ship is sold for scrap at $100 per ton. If the shipping company uses MARR of 6%, what is the ship's lifetime worth?

3- A factory has to install a smoke reduction system or pay a fine of $10,000 per month. The design of the system is outsourced to an A&E firm for $60,000 and takes six months to complete. The assembly, installation, and test are to be done on site. The cost of material is $100,000, the cost of labor is $60,000, and it takes eight months to bring the system to the operational level and avoid the penalty payment. The system needs an equivalent of one person at $25,000 per year to operate. The life of the system is five years, and the cost of removal and salvage value are the same. What is the LTW? The payments are made at the end of performance. Do not consider tax and depreciation. Use a10% per year interest rate.

4- Using the estimates of Example 2.2 and assuming the cost of a luxury home per square foot is 1.5 times the value stated, calculate the lifetime worth of a 5,000-square-foot luxury house if the rent is $2,000 per month adjusted every year for an inflation rate of 3% for a 20-year life. Use a 6% interest rate.

5- Define parameters you would use to determine the LTW of

a.   A football player

b.   A basketball player

6- An aerospace company asks you to calculate the LTW of their new passenger aircraft. They have just finished their conceptual design. How do you start looking at this problem? What information do you need? Which method would you use? What parameters do you look at? Briefly explain.

7- In problem 3, how will tax affect the LTW? How about tax and inflation?

8- A city transport authority is conducting an economical analysis on opening a new bus route. The study produces the following data:

*   The purchase price of a 30-passenger bus is $100,000

*   The mean time between failures (MTBF) is 5 1/2 months and the mean time to repair

(MTTR) is 1/2 month.

*   The administration cost of the company allocated to each bus service is $50 per

month.

*   Each major overhaul done during the 1/2 month repair, costs $2,000. This overhaul

cost increases by $500 every year (i.e., $2,500 for in the second year and so on).

*   The bus ticket is $1 per person/service route of which $ 0.75 is the direct operating

cost. The bus goes through 20 service routes per working day with a full load on each route.

If we consider a month to be 30 days (360 days/year) and the life of every bus to be 15 years,

what is the LTW of this bus if the MARR for the transport authority is 10%?

CHAPTER 12

LTW of Software System

INTRODUCTION

The major difference in the estimation of software LTW with the type of systems we have been discussing so far is in the estimation of its development cost. Software development involves more than writing codes. It is a systematic process that analyzes the requirements, prototypes and designs alternative approaches, then implements, tests, and delivers an automated solution. Traditional software development is performed in stages, from requirements gathering and analysis, through customer acceptance, and transition to operations and maintenance. The overall process is defined as the software lifetime.

SOFTWARE DEVELOPMENT PROCESS

The traditional software engineering lifetime transitions through a sequence of activities that define the lifetime, with each sequence completed prior to beginning the next phase. The traditional approach to software development is described as a "waterfall" approach. A graphical representation of the process often looks like a series of descending steps, giving the process the appearance of water falling from pool to pool on the journey toward the basin.

The traditional software lifetime is divided into 7 distinctive phases. These phases more or less correspond to the lifetime phases of the generalized system discussed previously. They are as follows:

1.   Requirements Analysis

2.   Preliminary Design

3.   Detailed Design

4.   Implementation

5.   System Testing

6.   Acceptance Testing

7.   Operation and Maintenance

REQUIREMENTS ANALYSIS

Requirements Analysis may include the effort to gather user requirements, but is most often the period for the system developer to understand what the user needs or has requested through the system requirements. This phase attempts to eliminate any misunderstanding of expectations and allows for negotiation on requirements that will be too costly to implement as specified. Attempts at an estimation of system size or the cost of implementation have an uncertainty of greater than 75%. Approximately 6% of the cost of development efforts have been spent up to this point.

PRELIMINARY DESIGN

The Preliminary Design phase is the developer's first opportunity to define and then to propose concepts and alternatives that could meet the requirements. At the closure of this phase, the design alternatives and costs per alternative are estimated. The cost estimations are made using rough architectures and productivity estimates to roughly gauge the final cost of the system development. The estimate at this phase is most often based upon a rule-of-thumb for the number of Source Lines of Code (software instructions) per subsystem (5000-7000 SLOC). The number of hours of effort required per subsystem is converted to dollars (labor dollars per hour) to arrive at the cost estimate for the proposed system. At this point in the software development lifetime, an additional 8% of the cost of the development effort has been spent (14% total). Estimation uncertainty is typically 50%.

DETAILED DESIGN

The Detailed Design phase follows the Preliminary Design, and is the period of time where the most cost-effective solution for meeting the requirements is prototyped, designed, and documented. Each subsystem is designed in detail. Detailed design typically takes 16% of the total development effort to complete the design phase. The estimate of the cost to complete the system can now be accurately made, as each unit and its respective interfaces are known and defined. Typical estimation rules define 100-125 Lines of Code (LOC) for each subsystem nodule or unit. At this point in the development process, 30% of the total system development cost has been spent. Estimation uncertainty, following the detailed design, is approximately 30%.

IMPLEMENTATION

Following the formal design, the system developer begins the actual writing of the software instructions that will execute in the system to meet the requirements. The implementation phase typically takes about 45% of the total development effort. Measures of the Source Lines of Code (SLOC) are carefully taken to gauge progress against the estimates in order to manage resources in light of the estimation uncertainty. Although the actual SLOC is known at the completion of this phase, the estimate uncertainty is about 12% due to the uncertainty in the number of tests and corrections that will have to be made to remove "bugs" in the program instructions.

SYSTEM TEST AND ACCEPTANCE

Following Implementation, the software will be integrated and tested in its final system configuration. Up to this point in the process, 75% of the total development effort has been expended. Integration and system test will require an additional 20% to check and correct the errors caused during Implementation. An additional 5% will be needed to prepare for and conduct formal acceptance testing. The uncertainty at this time can be as high as 5%. If all goes well, the software system will then be ready for delivery. The software development lifetime is now complete.

SOFTWARE ESTIMATION

Software development managers generally have well-defined processes for estimating the eventual system size and the labor effort for projects after a design has been completed. As mentioned above, approximately 30% of the total cost has been spent before a good idea has been formed as to how much the total system will cost so that budget is committed for further development. Most estimates, however, prior to the system design phase, typically range from a 50% overestimate to a 200% underestimate, due in part to insufficient information and requirements instability. At this point in the process, the system is often underestimated, causing the appearance of a cost overrun. Inaccurate estimates cause projects to be canceled due to monies being diverted to another project that has exceeded its budget allocation. Unfortunately, the initial estimate is most likely wrong, as the requirements are not yet well understood, and the alternative design trade-off studies have yet to be completed. The problem is to identify a method to estimate the cost of a software system with reasonable accuracy considering the limited information available to the software manager at the time of the completion of the system requirements.

Industry guidelines and experience have proven that estimates of the cost of developing new systems are most accurate when based upon the previous experiences of other similar projects (comparative estimation). Data must be collected on the total size and expended number of hours for each developed system in order to establish a statistical baseline from which to substantiate the next system development estimate. Software development metrics will be needed to develop an accurate estimation algorithm for the software manager's local environment. A major software development project typically takes three to five years to complete.

Software system size, often referred to as Source Lines of Code (SLOC) or Delivered Source Instructions (DSI), has a direct correlation to the number of hours required to produce a given system, using a specific development methodology and a specific development language.

Some industry analysts consider SLOC any line of code that does not include a comment, regardless of the complexity or simplicity of the statement. Others consider SLOC based on the number of possible paths in a software module. A programmer can break a statement based on his or her programming method, thus increasing or decreasing the size of the SLOC.

In addition, depending on the development tool, one line of code may deliver more functions as compared to another development tool. Estimation of the SLOC may then have to be adjusted depending on the choice of the development tool.

Research continues on industry trends in measuring and estimating software system size. Many different estimation methods exist, each having various degrees of success. However, almost all of the current estimation models require that the design of the system be available before the system can be estimated. In addition, each of the industry models are influenced by the development environment for each organization, indicating that there will most likely never be an industry-wide algorithm. Each organization should develop its own software measuring process and customize the models based upon local variables. To provide accurate estimates earlier in the software development lifetime, an empirical model needs to be developed using local environment development metrics.

Recent studies have found that by analyzing and grouping requirements from system requirement documents, an alternative estimation model can be developed for the software manager's local environment.

The studies revealed that

1.   The increasing number of system requirements causes a corresponding increase in the number of lines of code written to implement the system.

2.   The number of lines of code, per a certain grouping of requirements, increases as the system size increases.

From these trends, an algorithm can be developed to estimate the software system size based upon the observed trends in the data. Figure 2.6 illustrates the result of the analysis of a group of NASA projects at the Goddard Space Flight Center (GFSC). The system requirements were ordered and analyzed for developing a model for providing earlier estimates based upon requirement counts. The resulting method for estimating system size, based upon the count of the top three levels of functional and performance requirements, provided a curve that can be used to indicate the final system size to within 30% of actual, 90% of the time, in the NASA/GSFC environment. The Manion-Ardalan estimation curve based on the research work of David Manion (Figure 12.1) permits NASA managers to determine the cost of a software system based upon the number of the top three level of requirements when following their system requirements document development standards. Two SLOC is equivalent to one hour of effort over the development lifetime, giving the NASA manager a quick method for estimating total cost.

It has to be noted that software design is a relatively new subject and is still in the development stage. New software languages and methodologies reduce the number of codes required to satisfy a requirement. Estimation of the efforts needed to satisfy a requirement is, therefore, dependent on a lot of parameters; hence, the experience of the system designer plays a great role in the accuracy of the estimation. The Manion-Ardalan estimation curve, when tailored to the software manager's local environment, can be used to effectively and accurately estimate system development costs on any project prior to the Design and Implementation phases.

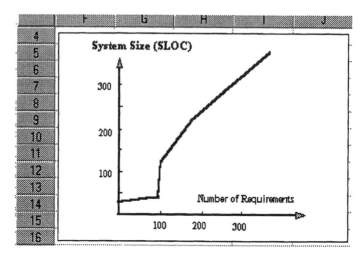

Figure 12.1

Manion-Ardalan Estimation curve

It is useful to note that the required lines of code for any specific program depend on the programming language. This has to be taken into account when estimating development cost of software.

PART 3

RETIREMENT AND REPLACEMENT

CHAPTER 13

Retirement and Replacement: Introduction and Definitions

INTRODUCTION

Depending on strategic business reasons, eventually a system is either retired from service or replaced by like or a new system. In this section, we first define retirement and replacement and discuss the reasons for retiring or replacing a system. Later, we will go through the financial analysis of the subject and use the tools of financial analysis in conjunction with the decision criteria to make a retirement or replacement decision.

RETIREMENT

Retirement is the action of removing a system from active service without employing another similar system to perform in its place. A system could be retired for two main reasons:

(1) The purpose for which the system was acquired is no longer valid. A helicopter is purchased to access a location where there were no roads; the road is built and the helicopter is no longer needed. The helicopter is then retired from service. It is sold or used for performing a different service. As far as our present project is concerned, the helicopter is retired, exactly as a person is retired or laid off from a company. He or she can and

probably will perform the same or different functions somewhere else, but is retired from the present service.

(2)   The investment in the system can bring higher returns if it is employed for another purpose. An old rental property is in bad shape. Its net present worth which includes the net of salvage value and the costs of its future operations and upkeep becomes unattractive when compared to employing the proceeds from its sale into better use in some other opportunity. The property is then sold, and the proceeds are invested elsewhere.

In both cases, the system is not being replaced, and the function it performed is no longer performed. In effect the "function" is retired.

Example 13.1

A property management company can keep operating an apartment house complex they own for five more years and receive $12M per year net of all the expenses. The resale value of the apartment house after five years is $22M. They also have the alternative opportunity of selling the apartment house for $20M now and investing the money at another business with an annual rate of return of 8%. Which alternative should they choose? Depreciation, inflation, and tax will not be considered.

Solution:

Note that what they paid for the apartment at the beginning is sunk cost and is not important at this moment. The cash flow diagram of the first alternative is shown below.

The NPW of this alternative at 8% is

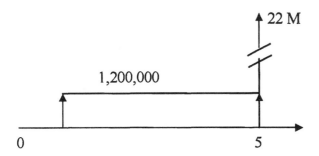

$$NPW = 1.2(P/A, 8, 5) + 22\ (P/F, 8, 5) = 19.76,\ \text{i.e.,}\ \$19.76M$$

This is less than the $20M they can obtain by selling the building now. They should

therefore sell the building and invest the proceeds at 8%.

OPPORTUNITY COST

In Example 13.1, if the company had decided to keep the building, they would have

lost the opportunity of investing the $20M they would have received by selling it. The $20M

is the cost of the lost opportunity. This value is therefore called the opportunity cost. We

can present another cash flow for Example 13.1 using the concept of opportunity cost and

can then calculate the NPW.

The NPW is

$$NPW = -20 -1.3(P/A, 8, 5) + 22\ (P/F, 8, 5) = .016$$

We reach the same conclusion, i.e., keeping the building for the next five years (the planning

horizon) has a negative NPW and, hence, it is not advisable.

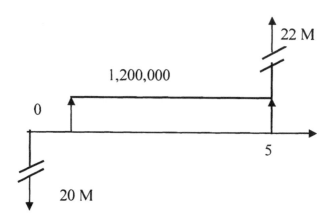

The retirement problem in effect comes down to checking the future prospects of the project against an investment opportunity with a fixed rate of return.

REPLACEMENT

The problem of replacement differs from retirement in that the need is still there for the function, but fulfilling the function with another system is more advantageous. Some of the reasons for which a system may be replaced are briefly mentioned here:

* End of physical life, i.e., no repair will bring it back to an acceptable performance level

* Lack of support by supplier

* High cost of maintenance

* High cost of operation

* Need for higher quality product that cannot be produced by the present system

* Need for higher productivity and/or efficiency

* Move to less labor-intensive systems

* Safety--the present system no longer satisfies safety standards or regulations

* New environmental regulations are imposed that the system cannot satisfy

* Added prestige of possessing a higher technology model

* Obsolescence (going out of use or fashion)

For any of the above reasons, the system is then replaced with a similar or different system to continue generally the same operation and deliver the output. The replaced system may or may not have a resale or salvage value, and the replacing system may or may not have the same lifetime financial profile.

PROBLEMS

1- Mr. Investgood owns a coin operating game machine he bought for $4,000 three years ago. He divides the $10 a day income from the machine with the owner of the arcade. The resale value of this machine is $500. The local bank has an investment opportunity for him with 8% annual return. The threat of a new computerized game will make his machine obsolete in six months. Should he retire his machine and go out of this business?

2- A plate forming machine was purchased four years ago at $100,000. The owner used a 10-year straight-line depreciation with an assumed resale value of $20,000 for this machine. He has an offer to sell this machine for $25,000. What is his opportunity cost?

3- You are using a car you bought three years ago for $20,000 only to go to work and back. A new metro station is to be opened one block from your house that will take you to a distance of one block from your office. Somebody suggests that you should sell your car and go to work by metro. What are the parameters you consider in deciding to accept the suggestion?

4- In the example of the helicopter and the road mentioned in the text, is the helicopter retired or replaced with the road? Discuss.

CHAPTER 14

Replacement Decision Making

REPLACEMENT PHILOSOPHY

All of the reasons listed for replacement in the last chapter can be grouped as economic mortality. That is, the need for the functions that the system performs still exists, but its net present worth for the rest of the system's intended life becomes less than that of a new system for the same time period. This occurs mainly because of the increase in the maintenance cost and the reduction in the productivity of the aging system. It also can occur because more efficient systems enter the marketplace. We have to make the choice--continue to maintain and operate the present system for the foreseeable future (planning horizon), or replace it with a new system. Financial analysis using a cost/time profile will be used as a decision-making tool.

The financial analysis of the replacement decision requires the knowledge of how the specific system characteristics that influence costs and benefits change with time. The future costs and benefits of keeping the present system, referred to as the defender, and the costs of purchasing, commissioning, and maintaining the replacement system, called the challenger, together with its associated benefits, have to be compared.

As the present system ages, it starts to deteriorate. Its resale value drops, its depreciation tax benefits may have ended, its operating characteristics change, and it generally becomes less

161

productive and more costly to operate and maintain. The challenger could be a system exactly like the defender, albeit a new one, that is, a new system with the same initial parameters as the defender had when it was installed. On the other hand, technologies involved in development, production, operation, and maintenance of systems change with time. Therefore, the challenger could be from new production lots of the same system or could be a completely new lot that basically satisfies the same needs but has different characteristics. The changes in system characteristics incorporated in new design contribute to the effectiveness and cost factors of the system during its life cycle. The new systems normally have a longer life span, are easier to maintain and operate, and are also more productive. Therefore, it may be advantageous to replace the system with a new one that has entered the market.

Industrial managers have always been concerned with the problem of system replacement. The first systematic approach to this problem was done by George Terborgh of the Machinery and Allied Product Institute who first coined the words defender and challenger. Since that time, others have approached this problem from different points of view, each emphasizing a specific area of concern.

Deciding when it is the proper time to replace an operating system with a new system depends on how age has affected the productivity and costs of the defender. It also depends on how product and process technological innovations have affected the various characteristics of the challengers entering the market. External conditions such as depreciation and tax also enter into the decision making process. There are other considerations in the areas of safety, environment, and even prestige that enter into the replacement decision. However, these can all be connected to profitability and cost factors. A replacement decision based on sound scientific reasoning can save time and effort; the wrong decision will waste money.

ECONOMIC LIFE AND CONTINUOUS REPLACEMENT

In Part 2 we discussed different definitions of life, among them the economic life. This is the length of time for which the profitability of the system is optimized if it is continuously replaced with a system of exactly the same cost and benefit parameters. If the system is kept in service one year longer or one year shorter, the profitability of the system is not maximized. This is based on the assumption that the net system profitability reduces with time, e.g., the cost of operation increases and the productivity falls. The resale value also reduces as the system ages. We will show that with these assumptions, if we calculate the EUAW and plot it against the years of keeping the system in operation, there is a specific lifetime that the EUAB is a maximum or the EUAC is a minimum. A typical cash flow diagram representing the aging system is shown in Fig. 14.1.

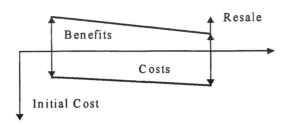

Fig. 14.1

Fig. 14.1 can be broken into three parts as shown in Fig. 14.2

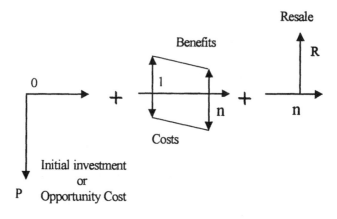

Fig. 14.2

To find the economic life, we have to follow the following steps:

<u>Step 1</u>:

Draw the cash flow diagram assuming a life of one year.

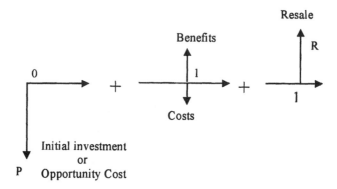

<u>Step 2</u>:

Calculate the EUAW for the three parts of a one-year life cash flow

EUAW (P) = P (A/P, i, 1)

EUAW (R) = R (A/F, i, 1)

EUAW (C or B) = [(B-C) (P/A, i, 1)] (A/P, i, 1)

<u>Step 3</u>:

Repeat steps 1 and 2 for every life year up to year n.

EUAW (P) = P (A/P, i, n)

EUAW (R) = R (A/F, i, n)

$$EUAW (B\text{-}C) = [\sum_{1}^{n} (B\text{-}C) (P/A, i, n)] (A/P, i, n)$$

The plot of the above three expressions is as presented in Fig. 14.3a. It is assumed, as usually is the case, that as years go by the benefits are reduced and the costs are increased.

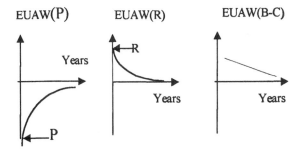

Fig. 14.3a

Step 4:

Plot the total EUAW against a horizontal axis representing the life years 1 to n. The

year representing the maximum or minimum is the economic life EL as shown in Fig. 14.3b.

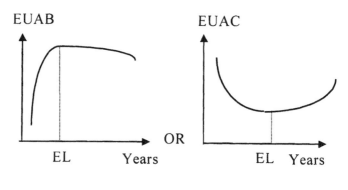

Fig. 14.3b

You should remember from Part 1 that an EUAW of A_n at any year n is the representation of

a A_n cash flow for all the years one to n and not the cash flow at year n. Therefore, the

optimum situation, i.e., maximum benefit or minimum cost, is obtained if the system is

replaced every EL years, EL being the economic life.

$$N_{optimum} = EL \qquad\qquad (14.1)$$

Example 14.1

Air handling equipment is bought to work with the air conditioning system of a

building. The initial cost is $18,000. The cost of maintenance and operation of this

equipment is $5,100 for the first year increasing by $1,800 every year. Resale value at the

end of the first year is $12,000 decreasing by $3,000 each year. The MARR for the owner of this air conditioning system is 20%. The pump output remains constant for 10 years. What is the economic life of this equipment? Do not consider tax and depreciation.

Solution:

Since the output of the pump is constant, the annual benefits from the operation remain constant and will not affect the economic life calculation. The economic life, therefore, is the year of life where the EUAC is a minimum.

One-Year Life

Step 1:

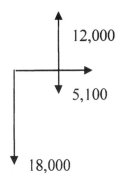

Step 2:

EUAC of initial cost = 18,000 (A/P,20, 1) = 21,600

EUAC of operation = 5,100 (A/F,20,1) = 5,100

EUAB of resale = 12,000 (A/F,20, 1) = 12,000

EUAC for one-year life = Total EUAB-Total EUAC = - 14,700

<u>Two-Year Life</u>

<u>Step 1 repeat</u>

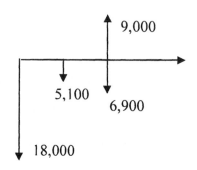

<u>Step 2 repeat</u>

EUAC of initial cost = - 18,000 (A/P,20, 2) = - 11,782

EUAC of operation = - [5,100 (P/F,20, 1) + 6,900 (P/F,20, 2)] (A/P, 20, 2) = - 5,917

EUAB of resale = 9,000 (A/F,20, 2) = 4,091

<u>EUAC for two-year life = - 13,608</u>

<u>Three-Year Life</u>

<u>Step 1 repeat</u>

<u>Step 2 repeat</u>

EUAC of initial cost = - 18,000 (A/P,20,3) = - 8545

EUAC of operation = - [5,100 (P/F,20, 1) + 6,900 (P/F,20, 2) + 8,700 (P/F,20, 3)] (A/P, 20,

3) = - 6,682

EUAB of resale = 6,000 (A/F,20, 3) = 1,648

<u>EUAC for three-year life = - 13,579</u>

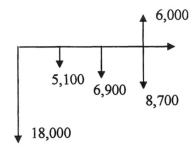

Four-Year Life

Step 1 repeat

EUAC of initial cost = - 18,000 (A/P,20,4) = -6,953

EUAC of operation = - [5,100 (P/F,20, 1) + 6,900 (P/F,20, 2) + 8,700 (P/F,20, 3)

+ 10,500 (P/F,20,4)] (A/P, 20, 4) = -7,394

EUAB of resale = 3,000 (A/F,20, 4) = 559

EUAC for four-year life = - 13,788

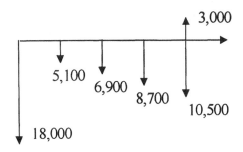

Five-Year Life

Step 1 repeat

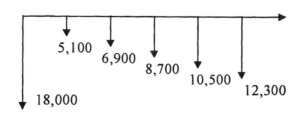

EUAC of initial cost = - 18,000 (A/P,20,5) = -6,019

EUAC of operation = - [5,100 (P/F,20, 1) + 6,900 (P/F,20, 2) +8,700 (P/F,20, 3)

+ 10,500 (P/F,20,4) + 12,300 (P/F,20, 5] (A/P, 20, 5) = -8,053

EUAB of resale = 0,000 (A/F,20, 4) = 0000

EUAC for five-year life = -14,072

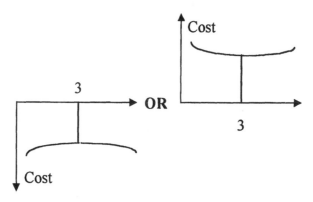

The EUAC plot is shown and the economic life is three years.

The above example demonstrates that if we continuously replace the system every three years, the financial effect is $13,579 per year. If we continuously replace the system for two or four years, the equivalent annual cost would be $13,608 and $13,788, respectively. Therefore, it is obvious that our best course of action is to replace the system with exactly the same system every three years.

COMPUTER DETERMINATION OF THE ECONOMIC LIFE

It is very easy to write a spreadsheet-based computer program to calculate the EUAW for any life and find the economic life. The program EL, based on the Quattro Pro spreadsheet (for Windows) will calculate and plot the EUAWs and determine the economic life.

A similar program for Excel can easily be developed. The inputs, the outputs, and the plots are shown here using Example 14.2.

Example 14.2

Clean Car Corporation (CCC), a car wash business in Vienna, Virginia, bought an automated car washing system for $500,000. The cost of operation of this system (labor, material, overhead, administration, etc.) is $100,000 per year increasing by $5,000 each year. The income from the operation of this system is $250,000 per year decreasing due to loss of productivity by 10% per year. The resale value of this system at the end of years 1 to 10 is $350,000, $250,000, $200,000, $170,000, and remains at $170,000 for the rest of the 10 years. If CCC has an 8% MARR, what is the optimum economical life of this system? De

not consider depreciation and tax.

Solution:

Step 1:

We start by calculating the EUAW if CCC keeps the system for only one year.

Net income = 250,000-100,000 = 150,000. The cash flow for this situation is

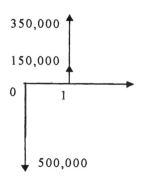

Step 2:

EUAC of initial cost = - 500000 (A/P, 8,1) = - 540000

EUAB of operation = 150000

EUAB of resale = 350000

EUAW for a one-year life = - 40000

Step 1 repeat

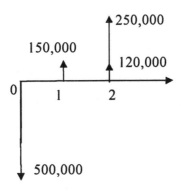

Step 2 repeat

EUAC of initial cost = - 500000 (A/P, 8, 2) = - 280400

EUAB of operation = [150000 (P/F, 8,1) + (250000 * 0.9 - 105000)(P/F, 8, 2)] (A/P, 8, 2) =

135580

EUAB of resale = 250000 (A/F, 8,2) = 120000

<u>EUAW for a two-year life = -24627</u>

We can of course continue doing this for 3, 4, 5 .... year life as we did for Example 14.1,

and obtain the economic life. An easier way is to use the already available spreadsheet

program EL. The solution is shown below.

Example 14.2

Investm 500000   Int. Rt.   0 08
         Tax Rt.   0 00
         Depreciation   N/A
         Type
         Resale   0
         Years

	Annual Financial Analysis							Capital Gain Analys					
Yr.	Income	Expense	Opr.Prf.	Depreci.	Taxable	Tax	IAT	Resale	Bk.Valu	Capt. G.	C.G.Tax	AT Rsl.	EUAW
1	250000	100000	150000		150000	0	150000	12000	500000	-488000	0	350000	-40000
2	225000	105000	120000		120000	0	120000	9000	500000	-491000	0	250000	-24615
3	202500	110000	92500		92500	0	92500	6000	500000	-494000	0	200000	-10102
4	182250	115000	67250		67250	0	67250	3000	500000	-497000	0	170000	-3145
5	164025	120000	44025		44025	0	44025	2000	500000	-498000	0	170000	2578
6	147623	125000	22623		22623	0	22622.5	1500	500000	-498500	0	170000	3456
7	132860	130000	2860.3	0	2860	0	2860.25	1000	500000	-499000	0	170000	1865
8	119574	135000	-15426	0	-15426	0	-15426	1000	500000	-499000	0	170000	-1039
9	107617	140000	-32383	0	-32383	0	-32383	1000	500000	-499000	0	170000	

EUAW (max) = 3456 at year 6

		EUAW					
Yea	Cash Fl	PW,1-r	Cash Fl	Salvage	Init. Co	TOTAL	LIFE
1	150000	138889	150000	350000	540000	-40000	0
2	120000	241770	135577	120192	280385	-24615	0
3	92500	315199	122308	61607	194017	-10102	0
4	67250	364630	110089	37727	150960	-3145	0
5	44025	394592	98828	28978	125228	2578	0
6	22623	408848	88440	23174	108158	3456	6
7	2860 3	410517	78849	19052	96036	1865	0
8	-15426	402183	69986	15983	87007	-1039	0
9							

Ec. Life   6

DEPRECIATION AND TAX

The effects of depreciation and tax in the replacement analysis are shown by way of the following example.

Example 14.3

Repeat Example 14.2 using straight-line depreciation, assuming a five-year life and no resale value for depreciation calculation. The aggregate tax rate for this case is 40%.

Solution:

Depreciation per year　　　D =(500000 - 0)/5 = 100000

Step 1:

Calculate the annual net income after tax (IAT) for year 1.

Year	Income	Expense	OP	Depr.	Taxable Income	Tax
	1	2	3=1-2	4	5=3-4	6=5*TR
1	250000	100000	150000	100000	50000	20000

This is the tax due to the operating income. Now we have to calculate the effect of capital

gain or loss.

Year	Resale	Book Value	Capital Gain/Loss	Tax On Cap G/L
	8	9	10=8-9	11
1	350000	400000	-50000	-20000

We now calculate the net income after tax.

Total Tax	IAT
12=11+6	3-12
0	150000

The cash flow then is

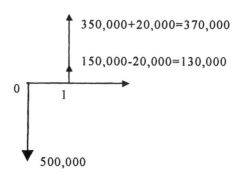

$350,000+20,000=370,000$

$150,000-20,000=130,000$

0

1

500,000

Therefore,

EUAC of initial cost = - 500000 (A/P, 8,1) = - 540000

EUAB of operation = 150000

EUAB of resale = 350000

EUAW for a one-year life = - 40000

<u>Step 2</u>:

Calculate the annual net income after tax (IAT) for years 1 and 2.

Year	Income	Expense	OP	Depr.	Taxable Income	Tax
	1	2	3=1-2	4	5=3-4	6=5*TR
1	250000	100000	150000	100000	50000	20000
2	225000	105000	120000	100000	20000	8000

This is the tax due to the operating income. Now we have to calculate the effect of capital

gain or loss.

Year	Resale	Book Value	Capital Gain/Loss	Tax On Cap G/L
	8	9	10=8-9	11
2	250000	300000	-50000	-20000

The cash flow then is

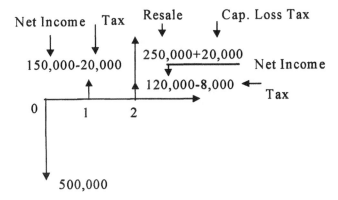

EUAC of initial cost = - 500000 (A/P, 8, 2) = - 280400

EUAB of operation = [130000 (P/F, 8,1) +112000 (P/F, 8, 2)] (A/P, 8, 2) =121348

EUAB of resale = 270000 (A/F, 8,2) = 129816

<u>EUAW for a two-year life = -29236</u>

We have to keep repeating the process until we get either a dip or an acme on the plot of

EUAW against the number of years. The following spreadsheet constructed with the same

procedure will provide us with the answer.

Example 14.3

Investm 500000   Int. Rt.   0.08
Tax Rt.   0.40
Depreciation   100000
Type   SL
Resale   0
Years   5

| Annual Financial Analysis | | | | | | | | Capital Gain Analysis | | | | |
Yr.	Income	Expense	Opr.Prf.	Depreci.	Taxable	Tax	IAT	Resale	Bk.Valu	Capt. G.	C.G.Tax	AT Rsl.	EUAW
1	250000	100000	150000	100000	50000	20000	130000	350000	400000	-50000	-20000	370000	-40000
2	225000	105000	120000	100000	20000	8000	112000	250000	300000	-50000	-20000	270000	-29231
3	202500	110000	92500	100000	-7500	-3000	95500	200000	200000	0	0	200000	-19025
4	182250	115000	67250	100000	-32750	-13100	80350	170000	100000	70000	28000	142000	-13394
5	164025	120000	44025	100000	-55975	-22390	66415	170000	0	170000	68000	102000	-8545
6	147623	125000	22622.5		22623	9049	13573.5	170000	0	170000	68000	102000	-6642
7	132860	130000	2860.25	0	2860	1144	1716.15	170000	0	170000	68000	102000	-6620
8	119574	135000	-15426	0	-15426	-6170	-9255.5	170000	0	170000	68000	102000	-7635
9	107617	140000	-32383	0	-32383	-12953	-19430	170000	0	170000	68000	102000	

EUAW MAX= -6620   At Yr. 7

| | | EUAW | | | | | |
Year	Cash Fl	PW,1-n	Cash F	Salvage	Init. Co	TOTAL	LIFE
1	130000	120370	130000	370000	540000	-40000	0
2	112000	216392	121346	129808	280385	-29231	0
3	95500	292203	113385	61607	194017	-19025	0
4	80350	351263	106054	31513	150960	-13394	0
5	66415	396464	99297	17387	125228	-8545	0
6	13574	405017	87612	13904	108158	-6642	0
7	1716.2	406019	77985	11431	96036	-6620	7
8	-9255	401018	69783	9590	87007	-7635	0
9							

Ec. Life   7

Note: Example 14.2 is a special version of Example 14.3, and the same results can be obtained by using a zero tax rate in the spreadsheet of Example 14.3.

The spreadsheet table looks very similar to the tables used for the manual calculations. In fact, the table can also be used for manual calculation by calculating and inserting the numbers rather than having the computer do it. Notice the PW, Yr.1-n column. This column is the NPW of all the annual cash flows from year 1 to the year n. The initial cost column under the

EUAW part of the table calculates the EUAW corresponding to the NPW for each year. Also notice that the introduction of tax into the game has resulted in less total income.

REPLACEMENT PROCESS

Previously, we discussed the reasons for replacement. Now we have to go through the process of assessing the economical justification for replacing a system. The central question is, should the system be replaced now? If not now, when? The decision to keep the defender or to replace it with the challenger needs the financial analysis of both of the alternatives. This requires a good picture of the costs and benefits involved in order to keep the system operational and productive. We have to compare this with the acquisition and operation costs and benefits of replacing it with a new system. As we discussed in the section on financial analysis (Part 1), in cases where the benefits are not exactly known, a subjective value or estimate has to be assigned.

The cash flow diagram of Fig. 14.4 shows all the benefits and costs of the defender from the time we are considering the replacement decision, year by year until the end of its physical life. At the time we are considering the possibility of replacement, the defender could be at any point in its life, i.e. 1,2, or n years old. The salvage value at the decision time would be an income opportunity lost if we do not sell the defender now. Therefore, it is shown as a cost of keeping the defender. The capital gain or loss with its tax effects, if we decide to sell now, is also either a cost or a benefit. The totality of these last two items is represented as the "opportunity cost" in Fig. 14.4.

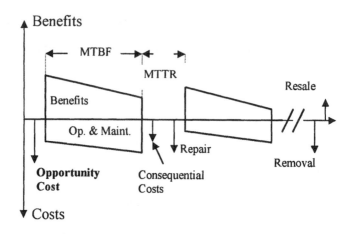

Fig. 14.4

The cash flow diagram of the challenger (see Fig 14.5) is rather similar, as it should be, to the defender's cash flow diagram. The initial investment represents all the costs of acquiring and putting the challenger into service, including initial inventory and training costs. In most cases, the challenger has better productivity, longer MTBF, shorter MTTR, and less operation and maintenance costs. This fact is reflected in the cash flow diagram of Fig. 14.5.

The decision to replace now or later can be made by analyzing the two cash flow diagrams.

The cash flow has to be constructed, and the EUAW of both systems has to be calculated year by year for each year from year n_0 onwards. For example, we should calculate the EUAW of the defender (the present system) and that of the challenger (the replacement candidate) for year n_0+1 (i.e., one year the preset time) and compare. This comparison tells us the benefit of keeping the defender for one more year versus selling the defender and purchasing the

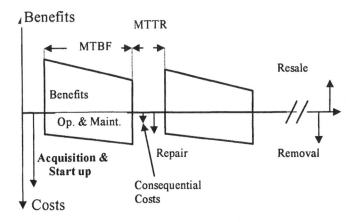

Fig. 14.5

challenger. If the resulting analysis favors the defender, then we carry the process one more year and compare the EUAW of the two alternatives. We continue the process until we arrive at our planning horizon year N_p. If at any year n_0 to N_p the financial analysis favors the challenger, we immediately replace the defender. Otherwise, we will keep the present system. If at any time in the future our planning horizon N_P changes, then we have to repeat the replacement analysis. The above analysis can best be understood through an example.

Example 14.4

The Owner of the Clean Car Corporation (Example 14.3) having operated his system for two years using SL depreciation is presented a new car wash system as a replacement for his present one. The new system has an initial cost of $600,000, including the cost of installation. The salesperson for the company producing a new system claims that the higher initial cost is compensated by higher productivity and lower operating cost. Moreover, he is willing to purchase the present system and move it out of the premises for $255,000. He

estimates that the income from the new system will be $300,000 in the first year (due to

higher initial productivity) reducing by 5% every year due to aging. The operating cost is

10% lower than the old system (due to more automation), increasing at a rate of 5% per year

as the system gets older. The resale value is estimated at $350,000 at the end of the first year,

reducing by $50,000 the next two years and staying constant at $170,000 thereafter.

Assuming the salesperson's assumptions are correct, how do you analyze this problem, and

what do you suggest to CCC. (The depreciation schedule remains the same.)

Solution:

Defender's Case:

We first have to calculate the opportunity cost of keeping the present system. The

parameters involved are

*The sales price of the old system, i.e., $ 255,000

*The capital gain tax, i.e., (sales price- book value at year 2) * Tax rate

= (255000- 300000)* 0.4 = -180000

*Therefore, the opportunity cost is   255000-(-18000) = 273000

From here on, the problems turn into a problem like Example 14.3 except that we have to

substitute the above opportunity cost for the initial cost and start from year 3 using the values

of its income, expense, depreciation, resale, etc. We therefore calculate and plot the EUAW

of the defender for one to n more years and compare this with that of the challenger.

Note that although the initial cost is replaced by the opportunity cost, the depreciation values are not changed and are based on the original cost.

Challenger's Case:

The challenger's case is exactly as the case of Example 14.3, albeit with new numbers representing the cost-benefit parameters of the new system. We can repeat the steps of Example 14.3 in the challenger's case. Rather than repeating those steps, we will use the spreadsheet to analyze this case. As we discussed before, the manual calculations can be made using the same table as in the spreadsheet. The computer spreadsheet is shown in its total form and broken down by the financial analysis of the defender and challenger. The analysis of the replacement situation is best done by observing the plot of the EUAWs of the defender and the challenger. We can analyze the relative value of the EUAWs against our planning horizon. From the plots shown in Fig. 14.6, we can see that if we are only thinking about the next three years, it is advisable to keep the defender. If our planning horizon is more than four years it is advisable to replace the defender by the challenger right now. Of course, the challenger then becomes a defender and its economics have to defend against a new challenger that may be introduced into the market. The matrix of Table 14.1 presents the defender-challenger economic analysis based on the calculation of the EUAW of the challenger and the defender.

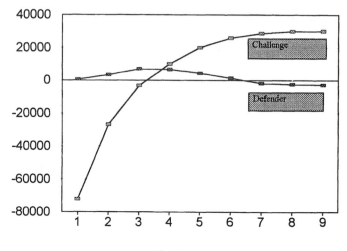

Fig. 14.1

Planning Horizon	1	2	3	4	5	6	7	8
Decision	K	K	K	R	R	R	R	R

Table 14.1

R: Replace by the challenger   K: Keep the old equipment

The spreadsheet calculation of the defender-challenger problem is shown in the next page.

CHALLENGER

Cost 600000	Int. Rate	0.08
	Tax	0.40
	Depreciation	120000
	Type	Str. Line
	Resale	0
	Years	5

Annual Financial Analysis / Capital Gain Analysis

Yr	Income	Expense	Opr.Prf	Depreciation	Taxable Inc	Tax	IAT	Resale	Book V	Cap.G	C.G Tax	IAT Resale	EUAW
1	300000	90000	210000	120000	90000	36000	174000	350000	480000	-130000	-52000	402000	72000
2	285000	94500	190500	120000	70500	28200	162300	250000	360000	-110000	-44000	294000	26740
3	270750	99225	171525	120000	51525	20610	150915	200000	240000	-40000	-16000	216000	3288
4	257213	104186	153026	120000	33026	13211	139816	170000	120000	50000	20000	150000	9988
5	244352	109396	134956	120000	14956	5983	128974	170000	0	170000	68000	102000	20043
6	232134	114865	117269		117269	46908	70361	170000	0	170000	68000	102000	25789
7	220528	120609	99919		99919	39968	59951	170000	0	170000	68000	102000	28704
8	209501	126639	82862		82862	33145	49717	170000	0	170000	68000	102000	29912
9	199026	132971	66055		6055	2422	63633	170000	-60000	230000	92000	78000	30037

EUAW MAX=30037 At Yr. 9

EUAW

Yr	Cash Flo	PW,Yr 1 - n	Cash Flo	Salvage	Initial. Cost	TOTAL	LIFE
1	174000	161111	174000	402000	648000	-72000	0
2	162300	300257	168375	141346	336462	-26740	0
3	150915	420058	162997	66535	232820	-3288	0
4	139816	522827	157852	33288	181152	9988	0
5	128974	610605	152930	17387	150274	20043	0
6	70361	654944	141674	13904	129789	25789	0
7	59951	689925	132516	11431	115243	28704	0
8	49717	716786	124731	9590	104409	29912	0
9	63633	748618	119839	6246	96048	30037	9

Econ Life 9

DEFENDER

Opportunity Cost	273000	Int. Rate	0.08
Market Price	255000	Tax Rate	0.40
Book Value	300000	Depreciation	100000
Cap. Gain Tax	-18000	Type	Str. Line
		Resale	0
		Years	5

Annual Financial Analysis / Capital Gain Analysis

Yr	Income	Expense	Opr.Prf	Depreciat	Taxable In	Tax	IAT	Resale	Book V	Cap.G	C.G Tax	IAT Resale	EUAW
1	202500	110000	92500	100000	-7500	-3000	95500	200000	200000	0	0	200000	660
2	182250	115000	67250	100000	-32750	-13100	80350	170000	100000	70000	28000	142000	3396
3	164025	120000	44025	100000	-55975	-22390	66415	170000	0	170000	68000	102000	6987
4	147623	125000	22623		22623	9049	13574	170000	0	170000	68000	102000	6638
5	132860	130000	2860		2860	1144	1716	170000	0	170000	68000	102000	4408
6	119574	135000	-15426		-15426	-6170	-9255	170000	0	170000	68000	102000	1433
7	107617	140000	-32383		-32383	-12953	-19430	170000	0	170000	68000	102000	-1820
8	96855	145000	-48145		-48145	-19258	-28887	170000	0	170000	68000	102000	-2416
9	87170	150000	-62830		-62830	-25132	-37698	170000	0	170000	68000	102000	-2876

EUAW MAX=6987 At Yr. 3

EUAW

Yr	Cash Flow	PW,Yr 1 - n	Cash Flo	Salvage	Initial. C	TOTAL	LIFE
1	95500	88426	95500	200000	294840	660	0
2	80350	157313	88216	68269	153090	3396	0
3	66415	210035	81501	31419	105933	6987	3
4	13574	220012	66426	22636	82424	6638	0
5	1716	221180	55396	17387	68375	4408	0
6	-9255	215348	46583	13904	59054	1433	0
7	-19430	204011	39185	11431	52436	-1820	0
8	0	204011	35501	9590	47506	-2416	0
9	0	204011	32658	8168	43702	-2876	0

Econ Life 3

185

MORE ABOUT REPLACEMENT

In reality, the values of parameters in the cash flow diagrams are time dependent, for example, the resale value of the defender in each year from n_0+n is different. The same is true for all other costs and benefits. The nature of the time dependency for all the parameters affecting costs and benefits, i.e., $f(t)$, is to be determined in each case when a replacement is considered. The outside factors such as tax and interest rate also have to be known. If the function $f(t)$ for each of these elements can be determined from the statistical data, a computer program can be developed to perform the analysis and provide us with an answer. The program will allow us to perform the following financial analyses.

1.   Calculate the life cycle worth of a system or a project

2.   Compare two projects and determine which one is more economical

3.   Compare a defender and a challenger and determine whether or not the defender should be replaced

4.   Compare the defender and the challenger and determine when in the future the defender should be replaced

PRIOR KNOWLEDGE OF REPLACEMENT TIME

In the majority of complex capital intensive systems there are many arrangements that have to be made before the system can come on-line and become productive. Preparations for proper installation have to be made. Operation and maintenance crews have to be trained, new inventory has to be stocked, and interfaces with the other systems must be prepared.

These activities and arrangements need time. Prior knowledge of the replacement time will facilitate these processes and will save time and money. The appropriate determination of the point of time when the replacement has to be made is therefore an essential issue to be considered where continuous, uninterrupted operation of the system is of importance. Special use of the model mentioned above can provide a reasonable estimate of the future time of replacement.

PROBLEMS

1- Independent Speedy Printing (ISP) purchased a printing press for $12,000. The annual operating and maintenance costs of this equipment are $3,500, increasing by $1,000 per year as the printing press gets older. The resale value at the end of the first year is $8,000, reducing by $2,000 each year. If the cost of money is assumed to be 20%, what is the economic life of this press? Do not consider depreciation and tax effects.

Assume ISP uses straight-line depreciation, six-year life, and zero resale value for depreciation purposes. Its aggregate tax rate is 40%, and it has a MARR of 10%. What is the economic life of this press? ISP is profitable and can use the tax effect of depreciation.

2- Equipment is bought for $20,000; its annual operational costs for the first three years are $3,000, $6,000, and $9,000, respectively, and they will increase by $6,000 every year from then on. The resale value of this equipment at any year equals its removal cost. The MARR for the owner of this equipment is 15%. What is the economic life of this equipment? Ignore income tax and depreciation.

3- A local newspaper purchased a color photography and reproduction system for $28,000. The annual revenue from the work of this system is $20,000 for the first year, decreasing by $500 due to reduced productivity. Annual operating and maintenance costs are $10,000 for the first year, increasing by $300 per year. The resale value at the end of the first year is $18,000, reducing by $4,000 each year until year 4 and remaining at $6,000 thereafter. The newspaper company uses straight-line depreciation, a five-year life, and zero resale value for depreciation purposes. Its aggregate tax rate is 55%. The company is profitable and can use the tax effect of depreciation.

a.   If the cost of money is assumed to be 5%, what is the economic life of this system?

b.   If at the end of the fourth year of operation the company is offered a price of $12,000 for the system, what would be the opportunity cost of not accepting the offer?

4- The College Park Sheet Metal Works (CPSMW), a very profitable Maryland corporation, purchased a press for $18,000. The annual operating and maintenance costs of this equipment are $3,000 for the first year, $3,500 for the second year, $5,500 for the third year, and they increase by $2,500 every year after that. The machine produces a sale of $13,000 per year. The resale value at the end of the first year is $ 12,000 and it reduces by $3,000 each year. CPSMW uses straight-line depreciation, a five-year life, and zero resale value for depreciation purposes. If the cost of money is assumed to be 10%, and its tax rate is 30%, what is the economic life of this press?

5-Transport International Incorporated (TII) is purchasing a new heavy-duty tractor-trailer truck for $250,000 and installing some special equipment on it at a cost of $90,000. The total cost is depreciated by the MACRS schedule. The company anticipates income of $300,000 the first year, with a reduction of $20,000 each year from the operation of this system. Its cost of operation is estimated to be $110,000 in the first year, increasing by $18,000 each year due to heavy downtime and maintenance cost. The estimated resale value of this system is $200,000 at the end of the first year, reducing to $150,000 by the end of the second year, and reducing by $20,000 each year after. TIJ has a tax rate of 40% and is using a MARR of 8%. Calculate the economic life of this system.

6- Vienna Laundry Services (VLS) is investigating replacement of a laundry and dry cleaning operation they bought eight years ago for $60,000 and which is fully depreciated. Another company is willing to buy the operation for $25,000 if VLS performs a major overhaul before transferring the operation to them. This overhaul will cost VLS $ 5,000. If the system is not sold, the estimated resale values, maintenance, and removal costs are as shown in Table 1.

Table 1

Year	1	2	3	4	5	6	7	8	9
Maintenance	8000	8000	8000	10000	15000	15000	20000	20000	20000
Resale	17000	15000	10000	5000	2000				
Removal						2000	2000	2000	2000

A new system can be bought for $70,000 with an installation cost of $10,000. The estimated resale values, maintenance, and removal costs of the new system would be as shown in Table 2.

It can be assumed that the productivity and operational cost of both systems are going to be the same. Tax and depreciation are not going to be considered.

RETIREMENT AND REPLACEMENT

Table 2

Year	1	2	3	4	5	6	7	8	9
Maintenance	0	0	1000	3000	5000	7000	9000	11000	13000
Resale	73000	70000	66000	62000	58000	54000	50000	46000	42000
Removal	0	0	0	0	0	0	0	0	0

Analyze this problem for VLS and make clear recommendations. Use a MARR of 10%. Show all your cash flow diagrams and details of any calculations.

7- ACME Corporation is investigating replacement of a punch press they bought two years ago for $48,000. Another company is willing to buy the press for $30,000. If the press is not sold, it will have an operating profit of $50,000 for this year, decreasing by $2,000 each year thereafter. Its resale value is estimated to be $25,000 by the end of this year, decreasing by $1,000 each year thereafter. A new press can be bought for $60,000. Its operating profit is $65,000 the first year, decreasing by $5,000 each year after that. Its resale value at the end of the first year is $30,000 and drops by $4,000 per year. The company's MARR is 5%. Analyze this problem for ACME and make clear recommendations with respect to ACME's strategic planning horizon varying from one to four years. Assume that the company operates in a total tax free zone. Show all of your cash flow diagrams and details of any calculations.

8-On January 1, 199X, the management of Sport Shoes Incorporated (SSI) discusses whether they should keep the shoe-making machinery they bought three years ago or replace it with new automated machinery. The present system has a market price of $300,000. Its original

price was $500,000, and it was depreciated using the straight-line method (no salvage value

and a five-year assumed life). Its market price at the end of the first year (December

31,199X) is estimated to be reduced to $220,000 and reduces by $20,000 each year up to the

end of the third year and by $60,000 each year after that. The income from the sale of the

products of the present system for the next four years is $200,000/year and is reducing by

$50,000 every year thereafter. The total expense for each year is $70,000. The new system,

if bought, will have a purchase price of $500,000 using the same depreciation method as the

old system. The income from its operations will be $280,000 each year. The first-year total

expense is $100,000 increasing by $10,000 every year. The estimated resale value of this

system at the end of the first year is $350,000, dropping by $50,000 each year. The MARR

for SSI is 8%, and their tax rate is 25%. Analyze this replacement problem for a planning

horizon of one to five years and recommend a decision to the management of SSI.

9-Metal Fabrication Inc. (MFI) is investigating replacement of a roller they bought five years

ago which is fully depreciated. Another company is willing to buy the roller for $23,000

delivered at their factory. The delivery will cost $3,000. If the roller is not sold the

maintenance cost will be $9,500 for the next year, increasing by $100 each year thereafter. Its

resale value is estimated to be $17,000 by the end of the year and $14,000, $11,000, and

$7,000 each year thereafter. A new roller can be bought for $70,000 with an installation cost

of $10,000. It will have no maintenance cost for the first two years, $1, 000 in year 3, increasing by $2,000 each year thereafter. Its resale value at the end of the first year is

$73,000, at the end of second year is $70,000, at the end of third year is $66,000 and drops

by $4,000 per year thereafter.

It can be assumed that the productivity and operational costs of both systems are going to be

the same. Tax and depreciation are not going to be considered. Analyze this problem for MFI

and make clear recommendations. Use a MARR of 10%. Show all your cash flow diagrams

and details of any calculations.

Proof of the Formulae

FUTURE VALUE-COMPOUND INTEREST

Interest rate for period	$= i,$
Interest for one year	$= Ai$
Total money after one year	$= A+Ai=A(1+i)$
Interest in the second year is	$= [\,A(1+i)\,]\,i$

Therefore, total money at the end of the second year is

$$= A(1+i)+ [\,A(1+i)\,]\,i = A(1+i)^2$$

Therefore, at the end of the nth year, the total money, that is the future value of

investment, is $\qquad FV= A_n=A(1+i)^n \qquad\qquad$ Eq. 1

PRESENT VALUE

It is obvious from the above equation that A is the present value of an amount A_n.

Therefore,

Present Value $\qquad\qquad PV = A_n(1+i)^{-n}$

or $\qquad\qquad\qquad\quad PV = FV(1+i)^{-n} \qquad\qquad$ Eq. 2

PRESENT VALUE OF UNIFORM SERIES

Present Value $PV = A_1 + A_2 + A_3 + \ldots\ldots\ldots A_n$

Substituting Eq. 2, we obtain $PV = \dfrac{(1+i)^n - 1}{i(1+i)^n}$ Eq. 3

ANNUAL WORTH OF THE PRESENT VALUE

This is the transposition of Eq. 3, where A is the EUAW; therefore,

$$EUAW = PV \times \frac{i(1+i)^n}{(1+i)^n - 1}$$ Eq.4

FUTURE VALUE OF UNIFORM SERIES

From Eq. 1 we can write

$$FV = A(1+i)^{n-1} + A(1+i)^{n-2} + A(1+i)^{n-3} \ldots\ldots\ldots\ldots + A(1+i) + A$$

By multiplying both sides by $(1+i)$ and factorizing A we get

$$i\,FV = A(1+i)^n + A(1+i)^{n-1} + A(1+i)^{n-2} \ldots\ldots\ldots\ldots + A + 1$$

$$i\,FV = A\left[(1+i)^n - 1\right]$$

$$FV = \frac{(1+i)^n - 1}{i}$$ Eq.5

ANNUAL WORTH OF A FUTURE VALUE:

This can be obtained by transposing Eq. 5:

$$EUAW = FV\,(i)\left[(1+i)^n - 1\right]^{-1}$$ Eq. 6

COMPOUND INTEREST TABLES
I=0.005

n	P/F	F/P	P/A	A/P	F/A	A/F	P/G	A/G	F/G
1	0.995	1.005	0.995	1.005	1.000	1.000	-0.000	-0.000	-0.0
2	0.990	1.010	1.985	0.504	2.005	0.499	0.990	0.499	1.0
3	0.985	1.015	2.970	0.337	3.015	0.332	2.960	0.997	3.0
4	0.980	1.020	3.950	0.253	4.030	0.248	5.901	1.494	6.0
5	0.975	1.025	4.926	0.203	5.050	0.198	9.803	1.990	10.1
6	0.971	1.030	5.896	0.170	6.076	0.165	14.655	2.485	15.1
7	0.966	1.036	6.862	0.146	7.106	0.141	20.449	2.980	21.2
8	0.961	1.041	7.823	0.128	8.141	0.123	27.176	3.474	28.3
9	0.956	1.046	8.779	0.114	9.182	0.109	34.824	3.967	36.4
10	0.951	1.051	9.730	0.103	10.228	0.098	43.386	4.459	45.6
11	0.947	1.056	10.677	0.094	11.279	0.089	52.853	4.950	55.8
12	0.942	1.062	11.619	0.086	12.336	0.081	63.214	5.441	67.1
13	0.937	1.067	12.556	0.080	13.397	0.075	74.460	5.930	79.4
14	0.933	1.072	13.489	0.074	14.464	0.069	86.583	6.419	92.8
15	0.928	1.078	14.417	0.069	15.537	0.064	99.574	6.907	107.3
16	0.923	1.083	15.340	0.065	16.614	0.060	113.424	7.394	122.8
17	0.919	1.088	16.259	0.062	17.697	0.057	128.123	7.880	139.5
18	0.914	1.094	17.173	0.058	18.786	0.053	143.663	8.366	157.2
19	0.910	1.099	18.082	0.055	19.880	0.050	160.036	8.850	175.9
20	0.905	1.105	18.987	0.053	20.979	0.048	177.232	9.334	195.8
21	0.901	1.110	19.888	0.050	22.084	0.045	195.243	9.817	216.8
22	0.896	1.116	20.784	0.048	23.194	0.043	214.061	10.299	238.9
23	0.892	1.122	21.676	0.046	24.310	0.041	233.677	10.781	262.1
24	0.887	1.127	22.563	0.044	25.432	0.039	254.082	11.261	286.4
25	0.883	1.133	23.446	0.043	26.559	0.038	275.269	11.741	311.8
26	0.878	1.138	24.324	0.041	27.692	0.036	297.228	12.220	338.4
27	0.874	1.144	25.198	0.040	28.830	0.035	319.952	12.698	366.1
28	0.870	1.150	26.068	0.038	29.975	0.033	343.433	13.175	394.9
29	0.865	1.156	26.933	0.037	31.124	0.032	367.663	13.651	424.9
30	0.861	1.161	27.794	0.036	32.280	0.031	392.632	14.126	456.0
31	0.857	1.167	28.651	0.035	33.441	0.030	418.335	14.601	488.3
32	0.852	1.173	29.503	0.034	34.609	0.029	444.762	15.075	521.7
33	0.848	1.179	30.352	0.033	35.782	0.028	471.906	15.548	556.3
34	0.844	1.185	31.196	0.032	36.961	0.027	499.758	16.020	592.1
35	0.840	1.191	32.035	0.031	38.145	0.026	528.312	16.492	629.1
40	0.819	1.221	36.172	0.028	44.159	0.023	681.335	18.836	831.8
45	0.799	1.252	40.207	0.025	50.324	0.020	850.763	21.159	1064.8
50	0.779	1.283	44.143	0.023	56.645	0.018	1035.697	23.462	1329.0
60	0.741	1.349	51.726	0.019	69.770	0.014	1448.646	28.006	1954.0
70	0.705	1.418	58.939	0.017	83.566	0.012	1913.643	32.468	2713.2
80	0.671	1.490	65.802	0.015	98.068	0.010	2424.646	36.847	3613.5
90	0.638	1.567	72.331	0.014	113.311	0.009	2976.077	41.145	4662.2
100	0.607	1.647	78.543	0.013	129.334	0.008	3562.793	45.361	5866.7
180	0.407	2.454	118.504	0.008	290.819	0.003	9031.336	76.212	22163.7
240	0.302	3.310	139.581	0.007	462.041	0.002	13415.540	96.113	44408.2
360	0.166	6.023	166.792	0.006	1004.515	0.001	21403.304	******	128903.0

197

n	P/F	F/P	P/A	A/P	F/A	A/F	P/G	A/G	F/G
1	0.990	1.010	0.990	1.010	1.000	1.000	0.000	0.000	0.000
2	0.980	1.020	1.970	0.508	2.010	0.498	0.980	0.498	1.000
3	0.971	1.030	2.941	0.340	3.030	0.330	2.921	0.993	3.010
4	0.961	1.041	3.902	0.256	4.060	0.246	5.804	1.488	6.040
5	0.951	1.051	4.853	0.206	5.101	0.196	9.610	1.980	10.101
6	0.942	1.062	5.795	0.173	6.152	0.163	14.321	2.471	15.202
7	0.933	1.072	6.728	0.149	7.214	0.139	19.917	2.960	21.354
8	0.923	1.083	7.652	0.131	8.286	0.121	26.381	3.448	28.567
9	0.914	1.094	8.566	0.117	9.369	0.107	33.696	3.934	36.853
10	0.905	1.105	9.471	0.106	10.462	0.096	41.843	4.418	46.221
11	0.896	1.116	10.368	0.096	11.567	0.086	50.807	4.901	56.683
12	0.887	1.127	11.255	0.089	12.683	0.079	60.569	5.381	68.250
13	0.879	1.138	12.134	0.082	13.809	0.072	71.113	5.861	80.933
14	0.870	1.149	13.004	0.077	14.947	0.067	82.422	6.338	94.742
15	0.861	1.161	13.865	0.072	16.097	0.062	94.481	6.814	109.69
16	0.853	1.173	14.718	0.068	17.258	0.058	107.273	7.289	125.79
17	0.844	1.184	15.562	0.064	18.430	0.054	120.783	7.761	143.04
18	0.836	1.196	16.398	0.061	19.615	0.051	134.996	8.232	161.47
19	0.828	1.208	17.226	0.058	20.811	0.048	149.895	8.702	181.09
20	0.820	1.220	18.046	0.055	22.019	0.045	165.466	9.169	201.90
21	0.811	1.232	18.857	0.053	23.239	0.043	181.695	9.635	223.92
22	0.803	1.245	19.660	0.051	24.472	0.041	198.566	10.100	247.16
23	0.795	1.257	20.456	0.049	25.716	0.039	216.066	10.563	271.63
24	0.788	1.270	21.243	0.047	26.973	0.037	234.180	11.024	297.35
25	0.780	1.282	22.023	0.045	28.243	0.035	252.894	11.483	324.32
26	0.772	1.295	22.795	0.044	29.526	0.034	272.196	11.941	352.56
27	0.764	1.308	23.560	0.042	30.821	0.032	292.070	12.397	382.09
28	0.757	1.321	24.316	0.041	32.129	0.031	312.505	12.852	412.91
29	0.749	1.335	25.066	0.040	33.450	0.030	333.486	13.304	445.04
30	0.742	1.348	25.808	0.039	34.785	0.029	355.002	13.756	478.49
31	0.735	1.361	26.542	0.038	36.133	0.028	377.039	14.205	513.27
32	0.727	1.375	27.270	0.037	37.494	0.027	399.586	14.653	549.41
33	0.720	1.389	27.990	0.036	38.869	0.026	422.629	15.099	586.90
34	0.713	1.403	28.703	0.035	40.258	0.025	446.157	15.544	625.77
35	0.706	1.417	29.409	0.034	41.660	0.024	470.158	15.987	666.03
40	0.672	1.489	32.835	0.030	48.886	0.020	596.856	18.178	888.64
45	0.639	1.565	36.095	0.028	56.481	0.018	733.704	20.327	1148.1
50	0.608	1.645	39.196	0.026	64.463	0.016	879.418	22.436	1446.3
60	0.550	1.817	44.955	0.022	81.670	0.012	1192.806	26.533	2167.0
70	0.498	2.007	50.169	0.020	100.676	0.010	1528.647	30.470	3067.6
80	0.451	2.217	54.888	0.018	121.672	0.008	1879.877	34.249	4167.2
90	0.408	2.449	59.161	0.017	144.863	0.007	2240.567	37.872	5486.3
100	0.370	2.705	63.029	0.016	170.481	0.006	2605.776	41.343	7048.1
180	0.167	5.996	83.322	0.012	499.580	0.002	5330.066	63.970	31958
240	0.092	10.893	90.819	0.011	989.255	0.001	6878.602	75.739	74926
360	0.028	35.950	97.218	0.010	3494.964	0.000	8720.432	89.699	313496

198

$I=0.02$

n	P/F	F/P	P/A	A/P	F/A	A/F	P/G	0.000	F/G
1	0.980	1.020	0.980	1.020	1.000	1.000	0.000	0.000	0.000
2	0.961	1.040	1.942	0.515	2.020	0.495	0.961	0.495	1.000
3	0.942	1.061	2.884	0.347	3.060	0.327	2.846	0.987	3.020
4	0.924	1.082	3.808	0.263	4.122	0.243	5.617	1.475	6.080
5	0.906	1.104	4.713	0.212	5.204	0.192	9.240	1.960	10.202
6	0.888	1.126	5.601	0.179	6.308	0.159	13.680	2.442	15.406
7	0.871	1.149	6.472	0.155	7.434	0.135	18.903	2.921	21.714
8	0.853	1.172	7.325	0.137	8.583	0.117	24.878	3.396	29.148
9	0.837	1.195	8.162	0.123	9.755	0.103	31.572	3.868	37.731
10	0.820	1.219	8.983	0.111	10.950	0.091	38.955	4.337	47.486
11	0.804	1.243	9.787	0.102	12.169	0.082	46.998	4.802	58.436
12	0.788	1.268	10.575	0.095	13.412	0.075	55.671	5.264	70.604
13	0.773	1.294	11.348	0.088	14.680	0.068	64.948	5.723	84.017
14	0.758	1.319	12.106	0.083	15.974	0.063	74.800	6.179	98.697
15	0.743	1.346	12.849	0.078	17.293	0.058	85.202	6.631	114.67
16	0.728	1.373	13.578	0.074	18.639	0.054	96.129	7.080	131.96
17	0.714	1.400	14.292	0.070	20.012	0.050	107.555	7.526	150.60
18	0.700	1.428	14.992	0.067	21.412	0.047	119.458	7.968	170.62
19	0.686	1.457	15.678	0.064	22.841	0.044	131.814	8.407	192.03
20	0.673	1.486	16.351	0.061	24.297	0.041	144.600	8.843	214.87
21	0.660	1.516	17.011	0.059	25.783	0.039	157.796	9.276	239.17
22	0.647	1.546	17.658	0.057	27.299	0.037	171.379	9.705	264.95
23	0.634	1.577	18.292	0.055	28.845	0.035	185.331	10.132	292.25
24	0.622	1.608	18.914	0.053	30.422	0.033	199.630	10.555	321.09
25	0.610	1.641	19.523	0.051	32.030	0.031	214.259	10.974	351.51
26	0.598	1.673	20.121	0.050	33.671	0.030	229.199	11.391	383.55
27	0.586	1.707	20.707	0.048	35.344	0.028	244.431	11.804	417.22
28	0.574	1.741	21.281	0.047	37.051	0.027	259.939	12.214	452.56
29	0.563	1.776	21.844	0.046	38.792	0.026	275.706	12.621	489.61
30	0.552	1.811	22.396	0.045	40.568	0.025	291.716	13.025	528.40
31	0.541	1.848	22.938	0.044	42.379	0.024	307.954	13.426	568.97
32	0.531	1.885	23.468	0.043	44.227	0.023	324.403	13.823	611.35
33	0.520	1.922	23.989	0.042	46.112	0.022	341.051	14.217	655.58
34	0.510	1.961	24.499	0.041	48.034	0.021	357.882	14.608	701.69
35	0.500	2.000	24.999	0.040	49.994	0.020	374.883	14.996	749.72
40	0.453	2.208	27.355	0.037	60.402	0.017	461.993	16.889	1020.1
45	0.410	2.438	29.490	0.034	71.893	0.014	551.565	18.703	1344.6
50	0.372	2.692	31.424	0.032	84.579	0.012	642.361	20.442	1729.0
60	0.305	3.281	34.761	0.029	114.052	0.009	823.698	23.696	2702.6
70	0.250	4.000	37.499	0.027	149.978	0.007	999.834	26.663	3998.9
80	0.205	4.875	39.745	0.025	193.772	0.005	1166.787	29.357	5688.6
90	0.168	5.943	41.587	0.024	247.157	0.004	1322.170	31.793	7857.8
100	0.138	7.245	43.098	0.023	312.232	0.003	1464.753	33.986	10612
180	0.028	35.321	48.584	0.021	1716.0	0.001	2174.413	44.755	76802
240	0.009	115.89	49.569	0.020	5744.4	0.000	2374.880	47.911	275222
360	0.001	1247.56	49.960	0.020	62328.1	0.000	2483.568	49.711	3098403

I = 0.03

n	P/F	F/P	P/A	A/P	F/A	A/F	P/G	0.000	F/G
1	0.971	1.030	0.971	1.030	1.000	1.000	0.000	0.000	0.0
2	0.943	1.061	1.913	0.523	2.030	0.493	0.943	0.493	1.0
3	0.915	1.093	2.829	0.354	3.091	0.324	2.773	0.980	3.0
4	0.888	1.126	3.717	0.269	4.184	0.239	5.438	1.463	6.1
5	0.863	1.159	4.580	0.218	5.309	0.188	8.889	1.941	10.3
6	0.837	1.194	5.417	0.185	6.468	0.155	13.076	2.414	15.6
7	0.813	1.230	6.230	0.161	7.662	0.131	17.955	2.882	22.1
8	0.789	1.267	7.020	0.142	8.892	0.112	23.481	3.345	29.7
9	0.766	1.305	7.786	0.128	10.159	0.098	29.612	3.803	38.6
10	0.744	1.344	8.530	0.117	11.464	0.087	36.309	4.256	48.8
11	0.722	1.384	9.253	0.108	12.808	0.078	43.533	4.705	60.3
12	0.701	1.426	9.954	0.100	14.192	0.070	51.248	5.148	73.1
13	0.681	1.469	10.635	0.094	15.618	0.064	59.420	5.587	87.3
14	0.661	1.513	11.296	0.089	17.086	0.059	68.014	6.021	102.9
15	0.642	1.558	11.938	0.084	18.599	0.054	77.000	6.450	120.0
16	0.623	1.605	12.561	0.080	20.157	0.050	86.348	6.874	138.6
17	0.605	1.653	13.166	0.076	21.762	0.046	96.028	7.294	158.7
18	0.587	1.702	13.754	0.073	23.414	0.043	106.014	7.708	180.5
19	0.570	1.754	14.324	0.070	25.117	0.040	116.279	8.118	203.9
20	0.554	1.806	14.877	0.067	26.870	0.037	126.799	8.523	229.0
21	0.538	1.860	15.415	0.065	28.676	0.035	137.550	8.923	255.9
22	0.522	1.916	15.937	0.063	30.537	0.033	148.509	9.319	284.6
23	0.507	1.974	16.444	0.061	32.453	0.031	159.657	9.709	315.1
24	0.492	2.033	16.936	0.059	34.426	0.029	170.971	10.095	347.5
25	0.478	2.094	17.413	0.057	36.459	0.027	182.434	10.477	382.0
26	0.464	2.157	17.877	0.056	38.553	0.026	194.026	10.853	418.4
27	0.450	2.221	18.327	0.055	40.710	0.025	205.731	11.226	457.0
28	0.437	2.288	18.764	0.053	42.931	0.023	217.532	11.593	497.7
29	0.424	2.357	19.188	0.052	45.219	0.022	229.414	11.956	540.6
30	0.412	2.427	19.600	0.051	47.575	0.021	241.361	12.314	585.8
31	0.400	2.500	20.000	0.050	50.003	0.020	253.361	12.668	633.4
32	0.388	2.575	20.389	0.049	52.503	0.019	265.399	13.017	683.4
33	0.377	2.652	20.766	0.048	55.078	0.018	277.464	13.362	735.9
34	0.366	2.732	21.132	0.047	57.730	0.017	289.544	13.702	791.0
35	0.355	2.814	21.487	0.047	60.462	0.017	301.627	14.037	848.7
40	0.307	3.262	23.115	0.043	75.401	0.013	361.750	15.650	1180.0
45	0.264	3.782	24.519	0.041	92.720	0.011	420.632	17.156	1590.7
50	0.228	4.384	25.730	0.039	112.797	0.009	477.480	18.558	2093.2
60	0.170	5.892	27.676	0.036	163.053	0.006	583.053	21.067	3435.1
70	0.126	7.918	29.123	0.034	230.594	0.004	676.087	23.215	5353.1
80	0.094	10.641	30.201	0.033	321.363	0.003	756.087	25.035	8045.4
90	0.070	14.300	31.002	0.032	443.349	0.002	823.630	26.567	11778.3
100	0.052	19.219	31.599	0.032	607.288	0.002	879.854	27.844	16909.6
180	0.005	204.503	33.170	0.030	6783.4	0.000	1076.339	32.449	220114.8
240	0.001	1204.85	33.306	0.030	40128.4	0.000	1103.549	33.134	1329614.0
360	0.000	41821.62	33.333	0.030	1394021	0.000	1110.798	33.325	46455360

n	P/F	F/P	P/A	A/P	F/A	A/F	P/G	0.000	F/G
1	0.962	1.040	0.962	1.040	1.000	1.000	0.000	0.000	0.0
2	0.925	1.082	1.886	0.530	2.040	0.490	0.925	0.490	1.0
3	0.889	1.125	2.775	0.360	3.122	0.320	2.703	0.974	3.0
4	0.855	1.170	3.630	0.275	4.246	0.235	5.267	1.451	6.2
5	0.822	1.217	4.452	0.225	5.416	0.185	8.555	1.922	10.4
6	0.790	1.265	5.242	0.191	6.633	0.151	12.506	2.386	15.8
7	0.760	1.316	6.002	0.167	7.898	0.127	17.066	2.843	22.5
8	0.731	1.369	6.733	0.149	9.214	0.109	22.181	3.294	30.4
9	0.703	1.423	7.435	0.134	10.583	0.094	27.801	3.739	39.6
10	0.676	1.480	8.111	0.123	12.006	0.083	33.881	4.177	50.2
11	0.650	1.539	8.760	0.114	13.486	0.074	40.377	4.609	62.2
12	0.625	1.601	9.385	0.107	15.026	0.067	47.248	5.034	75.6
13	0.601	1.665	9.986	0.100	16.627	0.060	54.455	5.453	90.7
14	0.577	1.732	10.563	0.095	18.292	0.055	61.962	5.866	107.3
15	0.555	1.801	11.118	0.090	20.024	0.050	69.735	6.272	125.6
16	0.534	1.873	11.652	0.086	21.825	0.046	77.744	6.672	145.6
17	0.513	1.948	12.166	0.082	23.698	0.042	85.958	7.066	167.4
18	0.494	2.026	12.659	0.079	25.645	0.039	94.350	7.453	191.1
19	0.475	2.107	13.134	0.076	27.671	0.036	102.893	7.834	216.8
20	0.456	2.191	13.590	0.074	29.778	0.034	111.565	8.209	244.5
21	0.439	2.279	14.029	0.071	31.969	0.031	120.341	8.578	274.2
22	0.422	2.370	14.451	0.069	34.248	0.029	129.202	8.941	306.2
23	0.406	2.465	14.857	0.067	36.618	0.027	138.128	9.297	340.4
24	0.390	2.563	15.247	0.066	39.083	0.026	147.101	9.648	377.1
25	0.375	2.666	15.622	0.064	41.646	0.024	156.104	9.993	416.1
26	0.361	2.772	15.983	0.063	44.312	0.023	165.121	10.331	457.8
27	0.347	2.883	16.330	0.061	47.084	0.021	174.138	10.664	502.1
28	0.333	2.999	16.663	0.060	49.968	0.020	183.142	10.991	549.2
29	0.321	3.119	16.984	0.059	52.966	0.019	192.121	11.312	599.2
30	0.308	3.243	17.292	0.058	56.085	0.018	201.062	11.627	652.1
31	0.296	3.373	17.588	0.057	59.328	0.017	209.956	11.937	708.2
32	0.285	3.508	17.874	0.056	62.701	0.016	218.792	12.241	767.5
33	0.274	3.648	18.148	0.055	66.210	0.015	227.563	12.540	830.2
34	0.264	3.794	18.411	0.054	69.858	0.014	236.261	12.832	896.4
35	0.253	3.946	18.665	0.054	73.652	0.014	244.877	13.120	966.3
40	0.208	4.801	19.793	0.051	95.026	0.011	286.530	14.477	1375.6
45	0.171	5.841	20.720	0.048	121.029	0.008	325.403	15.705	1900.7
50	0.141	7.107	21.482	0.047	152.667	0.007	361.164	16.812	2566.7
60	0.095	10.520	22.623	0.044	237.991	0.004	422.997	18.697	4449.8
70	0.064	15.572	23.395	0.043	364.290	0.003	472.479	20.196	7357.3
80	0.043	23.050	23.915	0.042	551.245	0.002	511.116	21.372	11781.1
90	0.029	34.119	24.267	0.041	827.983	0.001	540.737	22.283	18449.6
100	0.020	50.505	24.505	0.041	1237.624	0.001	563.125	22.980	28440.6
180	0.001	1164.129	24.979	0.040	29078.2	0.000	620.598	24.845	722455.6
240	0.000	12246.2	24.998	0.040	306130.1	0.000	624.459	24.980	7647251.5
360	0.000	1355196.1	25.000	0.040	33879878	0.000	624.993	25.000	846987946

n	P/F	F/P	P/A	A/P	F/A	A/F	P/G	0.000	F/G
1	0.952	1.050	0.952	1.050	1.000	1.000	0.000	0.000	0.0
2	0.907	1.103	1.859	0.538	2.050	0.488	0.907	0.488	1.0
3	0.864	1.158	2.723	0.367	3.153	0.317	2.635	0.967	3.1
4	0.823	1.216	3.546	0.282	4.310	0.232	5.103	1.439	6.2
5	0.784	1.276	4.329	0.231	5.526	0.181	8.237	1.903	10.5
6	0.746	1.340	5.076	0.197	6.802	0.147	11.968	2.358	16.0
7	0.711	1.407	5.786	0.173	8.142	0.123	16.232	2.805	22.8
8	0.677	1.477	6.463	0.155	9.549	0.105	20.970	3.245	31.0
9	0.645	1.551	7.108	0.141	11.027	0.091	26.127	3.676	40.5
10	0.614	1.629	7.722	0.130	12.578	0.080	31.652	4.099	51.6
11	0.585	1.710	8.306	0.120	14.207	0.070	37.499	4.514	64.1
12	0.557	1.796	8.863	0.113	15.917	0.063	43.624	4.922	78.3
13	0.530	1.886	9.394	0.106	17.713	0.056	49.988	5.322	94.3
14	0.505	1.980	9.899	0.101	19.599	0.051	56.554	5.713	112.0
15	0.481	2.079	10.380	0.096	21.579	0.046	63.288	6.097	131.6
16	0.458	2.183	10.838	0.092	23.657	0.042	70.160	6.474	153.1
17	0.436	2.292	11.274	0.089	25.840	0.039	77.140	6.842	176.8
18	0.416	2.407	11.690	0.086	28.132	0.036	84.204	7.203	202.6
19	0.396	2.527	12.085	0.083	30.539	0.033	91.328	7.557	230.8
20	0.377	2.653	12.462	0.080	33.066	0.030	98.488	7.903	261.3
21	0.359	2.786	12.821	0.078	35.719	0.028	105.667	8.242	294.4
22	0.342	2.925	13.163	0.076	38.505	0.026	112.846	8.573	330.1
23	0.326	3.072	13.489	0.074	41.430	0.024	120.009	8.897	368.6
24	0.310	3.225	13.799	0.072	44.502	0.022	127.140	9.214	410.0
25	0.295	3.386	14.094	0.071	47.727	0.021	134.228	9.524	454.5
26	0.281	3.556	14.375	0.070	51.113	0.020	141.259	9.827	502.3
27	0.268	3.733	14.643	0.068	54.669	0.018	148.223	10.122	553.4
28	0.255	3.920	14.898	0.067	58.403	0.017	155.110	10.411	608.1
29	0.243	4.116	15.141	0.066	62.323	0.016	161.913	10.694	666.5
30	0.231	4.322	15.372	0.065	66.439	0.015	168.623	10.969	728.8
31	0.220	4.538	15.593	0.064	70.761	0.014	175.233	11.238	795.2
32	0.210	4.765	15.803	0.063	75.299	0.013	181.739	11.501	866.0
33	0.200	5.003	16.003	0.062	80.064	0.012	188.135	11.757	941.3
34	0.190	5.253	16.193	0.062	85.067	0.012	194.417	12.006	1021.3
35	0.181	5.516	16.374	0.061	90.320	0.011	200.581	12.250	1106.4
40	0.142	7.040	17.159	0.058	120.800	0.008	229.545	13.377	1616.0
45	0.111	8.985	17.774	0.056	159.700	0.006	255.315	14.364	2294.0
50	0.087	11.467	18.256	0.055	209.348	0.005	277.915	15.223	3187.0
60	0.054	18.679	18.929	0.053	353.584	0.003	314.343	16.606	5871.7
70	0.033	30.426	19.343	0.052	588.529	0.002	340.841	17.621	10370.6
80	0.020	49.561	19.596	0.051	971.229	0.001	359.646	18.353	17824.6
90	0.012	80.730	19.752	0.051	1594.607	0.001	372.749	18.871	30092.1
100	0.008	131.501	19.848	0.050	2610.0	0.000	381.749	19.234	50200.5
180	0.000	6517.4	19.997	0.050	130327.8	0.000	399.386	19.972	2602956.7
240	0.000	121739.6	20.000	0.050	*********	0.000	399.957	19.998	*********
360	0.000	*********	20.000	0.050	*********	0.000	400.000	20.000	*********

I = 0.06

n	P/F	F/P	P/A	A/P	F/A	A/F	P/G	0.000	F/G
1	0.943	1.060	0.943	1.060	1.000	1.000	0.000	0.000	0.0
2	0.890	1.124	1.833	0.545	2.060	0.485	0.890	0.485	1.0
3	0.840	1.191	2.673	0.374	3.184	0.314	2.569	0.961	3.1
4	0.792	1.262	3.465	0.289	4.375	0.229	4.946	1.427	6.2
5	0.747	1.338	4.212	0.237	5.637	0.177	7.935	1.884	10.6
6	0.705	1.419	4.917	0.203	6.975	0.143	11.459	2.330	16.3
7	0.665	1.504	5.582	0.179	8.394	0.119	15.450	2.768	23.2
8	0.627	1.594	6.210	0.161	9.897	0.101	19.842	3.195	31.6
9	0.592	1.689	6.802	0.147	11.491	0.087	24.577	3.613	41.5
10	0.558	1.791	7.360	0.136	13.181	0.076	29.602	4.022	53.0
11	0.527	1.898	7.887	0.127	14.972	0.067	34.870	4.421	66.2
12	0.497	2.012	8.384	0.119	16.870	0.059	40.337	4.811	81.2
13	0.469	2.133	8.853	0.113	18.882	0.053	45.963	5.192	98.0
14	0.442	2.261	9.295	0.108	21.015	0.048	51.713	5.564	116.9
15	0.417	2.397	9.712	0.103	23.276	0.043	57.555	5.926	137.9
16	0.394	2.540	10.106	0.099	25.673	0.039	63.459	6.279	161.2
17	0.371	2.693	10.477	0.095	28.213	0.035	69.401	6.624	186.9
18	0.350	2.854	10.828	0.092	30.906	0.032	75.357	6.960	215.1
19	0.331	3.026	11.158	0.090	33.760	0.030	81.306	7.287	246.0
20	0.312	3.207	11.470	0.087	36.786	0.027	87.230	7.605	279.8
21	0.294	3.400	11.764	0.085	39.993	0.025	93.114	7.915	316.5
22	0.278	3.604	12.042	0.083	43.392	0.023	98.941	8.217	356.5
23	0.262	3.820	12.303	0.081	46.996	0.021	104.701	8.510	399.9
24	0.247	4.049	12.550	0.080	50.816	0.020	110.381	8.795	446.9
25	0.233	4.292	12.783	0.078	54.865	0.018	115.973	9.072	497.7
26	0.220	4.549	13.003	0.077	59.156	0.017	121.468	9.341	552.6
27	0.207	4.822	13.211	0.076	63.706	0.016	126.860	9.603	611.8
28	0.196	5.112	13.406	0.075	68.528	0.015	132.142	9.857	675.5
29	0.185	5.418	13.591	0.074	73.640	0.014	137.310	10.103	744.0
30	0.174	5.743	13.765	0.073	79.058	0.013	142.359	10.342	817.6
31	0.164	6.088	13.929	0.072	84.802	0.012	147.286	10.574	896.7
32	0.155	6.453	14.084	0.071	90.890	0.011	152.090	10.799	981.5
33	0.146	6.841	14.230	0.070	97.343	0.010	156.768	11.017	1072.4
34	0.138	7.251	14.368	0.070	104.184	0.010	161.319	11.228	1169.7
35	0.130	7.686	14.498	0.069	111.435	0.009	165.743	11.432	1273.9
40	0.097	10.286	15.046	0.066	154.762	0.006	185.957	12.359	1912.7
45	0.073	13.765	15.456	0.065	212.744	0.005	203.110	13.141	2795.7
50	0.054	18.420	15.762	0.063	290.336	0.003	217.457	13.796	4005.6
60	0.030	32.988	16.161	0.062	533.128	0.002	239.043	14.791	7885.5
70	0.017	59.076	16.385	0.061	967.932	0.001	253.327	15.461	14965.5
80	0.009	105.796	16.509	0.061	1746.600	0.001	262.549	15.903	27776.7
90	0.005	189.465	16.579	0.060	3141.075	0.000	268.395	16.189	50851.3
100	0.003	339.302	16.618	0.060	5638.368	0.000	272.047	16.371	92306.1
180	0.000	35897	16.666	0.060	598263	0.000	277.686	16.662	9968056
240	0.000	1184153	16.667	0.060	19735860	0.000	277.774	16.666	328926993
360	0.000	*********	16.667	0.060	********	0.000	277.778	16.667	*********

$I = 0.07$

n	P/F	F/P	P/A	A/P	F/A	A/F	P/G	0.000	F/G
1	0.935	1.070	0.935	1.070	1.000	1.000	0.000	0.000	0.0
2	0.873	1.145	1.808	0.553	2.070	0.483	0.873	0.483	1.0
3	0.816	1.225	2.624	0.381	3.215	0.311	2.506	0.955	3.1
4	0.763	1.311	3.387	0.295	4.440	0.225	4.795	1.416	6.3
5	0.713	1.403	4.100	0.244	5.751	0.174	7.647	1.865	10.7
6	0.666	1.501	4.767	0.210	7.153	0.140	10.978	2.303	16.5
7	0.623	1.606	5.389	0.186	8.654	0.116	14.715	2.730	23.6
8	0.582	1.718	5.971	0.167	10.260	0.097	18.789	3.147	32.3
9	0.544	1.838	6.515	0.153	11.978	0.083	23.140	3.552	42.5
10	0.508	1.967	7.024	0.142	13.816	0.072	27.716	3.946	54.5
11	0.475	2.105	7.499	0.133	15.784	0.063	32.466	4.330	68.3
12	0.444	2.252	7.943	0.126	17.888	0.056	37.351	4.703	84.1
13	0.415	2.410	8.358	0.120	20.141	0.050	42.330	5.065	102.0
14	0.388	2.579	8.745	0.114	22.550	0.044	47.372	5.417	122.1
15	0.362	2.759	9.108	0.110	25.129	0.040	52.446	5.758	144.7
16	0.339	2.952	9.447	0.106	27.888	0.036	57.527	6.090	169.8
17	0.317	3.159	9.763	0.102	30.840	0.032	62.592	6.411	197.7
18	0.296	3.380	10.059	0.099	33.999	0.029	67.622	6.722	228.6
19	0.277	3.617	10.336	0.097	37.379	0.027	72.599	7.024	262.6
20	0.258	3.870	10.594	0.094	40.995	0.024	77.509	7.316	299.9
21	0.242	4.141	10.836	0.092	44.865	0.022	82.339	7.599	340.9
22	0.226	4.430	11.061	0.090	49.006	0.020	87.079	7.872	385.8
23	0.211	4.741	11.272	0.089	53.436	0.019	91.720	8.137	434.8
24	0.197	5.072	11.469	0.087	58.177	0.017	96.255	8.392	488.2
25	0.184	5.427	11.654	0.086	63.249	0.016	100.7	8.639	546.4
26	0.172	5.807	11.826	0.085	68.676	0.015	105.0	8.877	609.7
27	0.161	6.214	11.987	0.083	74.484	0.013	109.2	9.107	678.3
28	0.150	6.649	12.137	0.082	80.698	0.012	113.2	9.329	752.8
29	0.141	7.114	12.278	0.081	87.347	0.011	117.2	9.543	833.5
30	0.131	7.612	12.409	0.081	94.461	0.011	121.0	9.749	920.9
31	0.123	8.145	12.532	0.080	102.1	0.010	124.7	9.947	1015.3
32	0.115	8.715	12.647	0.079	110.2	0.009	128.2	10.138	1117.4
33	0.107	9.325	12.754	0.078	118.9	0.008	131.6	10.322	1227.6
34	0.100	9.978	12.854	0.078	128.3	0.008	135.0	10.499	1346.6
35	0.094	10.677	12.948	0.077	138.2	0.007	138.1	10.669	1474.8
40	0.067	14.974	13.332	0.075	199.6	0.005	152.3	11.423	2280.5
45	0.048	21.002	13.606	0.073	285.7	0.003	163.8	12.036	3439.3
50	0.034	29.457	13.801	0.072	406.5	0.002	172.9	12.529	5093.3
60	0.017	57.946	14.039	0.071	813.5	0.001	185.8	13.232	10764.6
70	0.009	113.989	14.160	0.071	1614.1	0.001	193.5	13.666	22059.1
80	0.004	224.234	14.222	0.070	3189.1	0.000	198.1	13.927	44415.2
90	0.002	441.103	14.253	0.070	6287.2	0.000	200.7	14.081	88531.2
100	0.001	867.716	14.269	0.070	12381.7	0.000	202.2	14.170	175452.3
180	0.000	194571.8	14.286	0.070	2779583	0.000	204.1	14.285	*********
240	0.000	*********	14.286	0.070	********	0.000	204.1	14.286	*********
360	0.000	*********	14.286	0.070	********	0.000	204.1	14.286	*********

I = 0.08

n	P/F	F/P	P/A	A/P	F/A	A/F	P/G	0.000	F/G
1	0.926	1.080	0.926	1.080	1.000	1.000	0.000	0.000	0.0
2	0.857	1.166	1.783	0.561	2.080	0.481	0.857	0.481	1.0
3	0.794	1.260	2.577	0.388	3.246	0.308	2.445	0.949	3.1
4	0.735	1.360	3.312	0.302	4.506	0.222	4.650	1.404	6.3
5	0.681	1.469	3.993	0.250	5.867	0.170	7.372	1.846	10.8
6	0.630	1.587	4.623	0.216	7.336	0.136	10.523	2.276	16.7
7	0.583	1.714	5.206	0.192	8.923	0.112	14.024	2.694	24.0
8	0.540	1.851	5.747	0.174	10.637	0.094	17.806	3.099	33.0
9	0.500	1.999	6.247	0.160	12.488	0.080	21.808	3.491	43.6
10	0.463	2.159	6.710	0.149	14.487	0.069	25.977	3.871	56.1
11	0.429	2.332	7.139	0.140	16.645	0.060	30.266	4.240	70.6
12	0.397	2.518	7.536	0.133	18.977	0.053	34.634	4.596	87.2
13	0.368	2.720	7.904	0.127	21.495	0.047	39.046	4.940	106.2
14	0.340	2.937	8.244	0.121	24.215	0.041	43.472	5.273	127.7
15	0.315	3.172	8.559	0.117	27.152	0.037	47.886	5.594	151.9
16	0.292	3.426	8.851	0.113	30.324	0.033	52.264	5.905	179.1
17	0.270	3.700	9.122	0.110	33.750	0.030	56.588	6.204	209.4
18	0.250	3.996	9.372	0.107	37.450	0.027	60.843	6.492	243.1
19	0.232	4.316	9.604	0.104	41.446	0.024	65.013	6.770	280.6
20	0.215	4.661	9.818	0.102	45.762	0.022	69.090	7.037	322.0
21	0.199	5.034	10.017	0.100	50.423	0.020	73.063	7.294	367.8
22	0.184	5.437	10.201	0.098	55.457	0.018	76.926	7.541	418.2
23	0.170	5.871	10.371	0.096	60.893	0.016	80.673	7.779	473.7
24	0.158	6.341	10.529	0.095	66.765	0.015	84.300	8.007	534.6
25	0.146	6.848	10.675	0.094	73.106	0.014	87.804	8.225	601.3
26	0.135	7.396	10.810	0.093	79.954	0.013	91.184	8.435	674.4
27	0.125	7.988	10.935	0.091	87.351	0.011	94.439	8.636	754.4
28	0.116	8.627	11.051	0.090	95.339	0.010	97.569	8.829	841.7
29	0.107	9.317	11.158	0.090	103.966	0.010	100.574	9.013	937.1
30	0.099	10.063	11.258	0.089	113.283	0.009	103.456	9.190	1041.0
31	0.092	10.868	11.350	0.088	123.346	0.008	106.216	9.358	1154.3
32	0.085	11.737	11.435	0.087	134.214	0.007	108.857	9.520	1277.7
33	0.079	12.676	11.514	0.087	145.951	0.007	111.382	9.674	1411.9
34	0.073	13.690	11.587	0.086	158.627	0.006	113.792	9.821	1557.8
35	0.068	14.785	11.655	0.086	172.317	0.006	116.092	9.961	1716.5
40	0.046	21.725	11.925	0.084	259.057	0.004	126.042	10.570	2738.2
45	0.031	31.920	12.108	0.083	386.506	0.003	133.733	11.045	4268.8
50	0.021	46.902	12.233	0.082	573.770	0.002	139.593	11.411	6547.1
60	0.010	101.257	12.377	0.081	1253.213	0.001	147.300	11.902	14915.2
70	0.005	218.606	12.443	0.080	2720.080	0.000	151.533	12.178	33126.0
80	0.002	471.955	12.474	0.080	5886.935	0.000	153.800	12.330	72586.7
90	0.001	1018.915	12.488	0.080	12724	0.000	154.993	12.412	157924.2
100	0.000	2199.761	12.494	0.080	27485	0.000	155.611	12.455	342306.4
180	0.000	1038188	12.500	0.080	12977337	0.000	156.248	12.500	*********
240	0.000	*********	12.500	0.080	********	0.000	156.250	12.500	*********
360	0.000	*********	12.500	0.080	********	0.000	156.250	12.500	*********

$I = 0.09$

n	P/F	F/P	P/A	A/P	F/A	A/F	P/G	0.000	F/G
1	0.917	1.090	0.917	1.090	1.000	1.000	0.000	0.000	0.0
2	0.842	1.188	1.759	0.568	2.090	0.478	0.082	0.047	0.097
3	0.772	1.295	2.531	0.395	3.278	0.305	0.231	0.091	0.299
4	0.708	1.412	3.240	0.309	4.573	0.219	0.434	0.134	0.612
5	0.650	1.539	3.890	0.257	5.985	0.167	0.680	0.175	1.046
6	0.596	1.677	4.486	0.223	7.523	0.133	0.960	0.214	1.610
7	0.547	1.828	5.033	0.199	9.200	0.109	1.266	0.252	2.314
8	0.502	1.993	5.535	0.181	11.028	0.091	1.592	0.288	3.172
9	0.460	2.172	5.995	0.167	13.021	0.077	1.931	0.322	4.195
10	0.422	2.367	6.418	0.156	15.193	0.066	2.280	0.355	5.398
11	0.388	2.580	6.805	0.147	17.560	0.057	2.634	0.387	6.797
12	0.356	2.813	7.161	0.140	20.141	0.050	2.990	0.418	8.410
13	0.326	3.066	7.487	0.134	22.953	0.044	3.345	0.447	10.254
14	0.299	3.342	7.786	0.128	26.019	0.038	3.696	0.475	12.352
15	0.275	3.642	8.061	0.124	29.361	0.034	4.043	0.502	14.725
16	0.252	3.970	8.313	0.120	33.003	0.030	4.382	0.527	17.398
17	0.231	4.328	8.544	0.117	36.974	0.027	4.713	0.552	20.398
18	0.212	4.717	8.756	0.114	41.301	0.024	5.036	0.575	23.755
19	0.194	5.142	8.950	0.112	46.018	0.022	5.348	0.598	27.500
20	0.178	5.604	9.129	0.110	51.160	0.020	5.651	0.619	31.669
21	0.164	6.109	9.292	0.108	56.765	0.018	5.942	0.639	36.299
22	0.150	6.659	9.442	0.106	62.873	0.016	6.223	0.659	41.433
23	0.138	7.258	9.580	0.104	69.532	0.014	6.492	0.678	47.116
24	0.126	7.911	9.707	0.103	76.790	0.013	6.750	0.695	53.397
25	0.116	8.623	9.823	0.102	84.701	0.012	6.996	0.712	60.331
26	0.106	9.399	9.929	0.101	93.324	0.011	7.232	0.728	67.975
27	0.098	10.245	10.027	0.100	102.723	0.010	7.457	0.744	76.394
28	0.090	11.167	10.116	0.099	112.968	0.009	7.671	0.758	85.659
29	0.082	12.172	10.198	0.098	124.135	0.008	7.874	0.772	95.844
30	0.075	13.268	10.274	0.097	136.308	0.007	8.067	0.785	107.034
31	0.069	14.462	10.343	0.097	149.575	0.007	8.251	0.798	119.318
32	0.063	15.763	10.406	0.096	164.037	0.006	8.424	0.810	132.795
33	0.058	17.182	10.464	0.096	179.800	0.006	8.589	0.821	147.573
34	0.053	18.728	10.518	0.095	196.982	0.005	8.744	0.831	163.769
35	0.049	20.414	10.567	0.095	215.711	0.005	8.892	0.841	181.511
40	0.032	31.409	10.757	0.093	337.882	0.003	9.511	0.884	298.738
45	0.021	48.327	10.881	0.092	525.859	0.002	9.969	0.916	481.756
50	0.013	74.358	10.962	0.091	815.084	0.001	10.302	0.940	766.011
60	0.006	176.031	11.048	0.091	1944.792	0.001	10.713	0.970	1885.756
70	0.002	416.730	11.084	0.090	4619.223	0.000	10.919	0.985	4550.206
80	0.001	986.552	11.100	0.090	10951	0.000	11.020	0.993	10872
90	0.000	2335.527	11.106	0.090	25939	0.000	11.068	0.997	25850
100	0.000	5529.041	11.109	0.090	61423	0.000	11.091	0.998	61324
180	0.000	5454684	11.111	0.090	60607594	0.000	11.111	1.000	60607415
240	0.000	*********	11.111	0.090	********	0.000	11.111	1.000	*********
360	0.000	*********	11.111	0.090	********	0.000	11.111	1.000	*********

206

I = 0.1

n	P/F	F/P	P/A	A/P	F/A	A/F	P/G	0.000	F/G
1	0.909	1.100	0.909	1.100	1.000	1.000	0.000	0.000	0.0
2	0.826	1.210	1.736	0.576	2.100	0.476	0.826	0.476	1.0
3	0.751	1.331	2.487	0.402	3.310	0.302	2.329	0.937	3.1
4	0.683	1.464	3.170	0.315	4.641	0.215	4.378	1.381	6.4
5	0.621	1.611	3.791	0.264	6.105	0.164	6.862	1.810	11.1
6	0.564	1.772	4.355	0.230	7.716	0.130	9.684	2.224	17.2
7	0.513	1.949	4.868	0.205	9.487	0.105	12.763	2.622	24.9
8	0.467	2.144	5.335	0.187	11.436	0.087	16.029	3.004	34.4
9	0.424	2.358	5.759	0.174	13.579	0.074	19.421	3.372	45.8
10	0.386	2.594	6.145	0.163	15.937	0.063	22.891	3.725	59.4
11	0.350	2.853	6.495	0.154	18.531	0.054	26.396	4.064	75.3
12	0.319	3.138	6.814	0.147	21.384	0.047	29.901	4.388	93.8
13	0.290	3.452	7.103	0.141	24.523	0.041	33.377	4.699	115.2
14	0.263	3.797	7.367	0.136	27.975	0.036	36.800	4.996	139.7
15	0.239	4.177	7.606	0.131	31.772	0.031	40.152	5.279	167.7
16	0.218	4.595	7.824	0.128	35.950	0.028	43.416	5.549	199.5
17	0.198	5.054	8.022	0.125	40.545	0.025	46.582	5.807	235.4
18	0.180	5.560	8.201	0.122	45.599	0.022	49.640	6.053	276.0
19	0.164	6.116	8.365	0.120	51.159	0.020	52.583	6.286	321.6
20	0.149	6.727	8.514	0.117	57.275	0.017	55.407	6.508	372.7
21	0.135	7.400	8.649	0.116	64.002	0.016	58.110	6.719	430.0
22	0.123	8.140	8.772	0.114	71.403	0.014	60.689	6.919	494.0
23	0.112	8.954	8.883	0.113	79.543	0.013	63.146	7.108	565.4
24	0.102	9.850	8.985	0.111	88.497	0.011	65.481	7.288	645.0
25	0.092	10.835	9.077	0.110	98.347	0.010	67.696	7.458	733.5
26	0.084	11.918	9.161	0.109	109.182	0.009	69.794	7.619	831.8
27	0.076	13.110	9.237	0.108	121.100	0.008	71.777	7.770	941.0
28	0.069	14.421	9.307	0.107	134.210	0.007	73.650	7.914	1062.1
29	0.063	15.863	9.370	0.107	148.631	0.007	75.415	8.049	1196.3
30	0.057	17.449	9.427	0.106	164.494	0.006	77.077	8.176	1344.9
31	0.052	19.194	9.479	0.105	181.943	0.005	78.640	8.296	1509.4
32	0.047	21.114	9.526	0.105	201.138	0.005	80.108	8.409	1691.4
33	0.043	23.225	9.569	0.104	222.252	0.004	81.486	8.515	1892.5
34	0.039	25.548	9.609	0.104	245.477	0.004	82.777	8.615	2114.8
35	0.036	28.102	9.644	0.104	271.024	0.004	83.987	8.709	2360.2
40	0.022	45.259	9.779	0.102	442.593	0.002	88.953	9.096	4025.9
45	0.014	72.890	9.863	0.101	718.905	0.001	92.454	9.374	6739.0
50	0.009	117.391	9.915	0.101	1163.909	0.001	94.889	9.570	11139.1
60	0.003	304.482	9.967	0.100	3034.816	0.000	97.701	9.802	29748.2
70	0.001	789.747	9.987	0.100	7887.470	0.000	98.987	9.911	78174.7
80	0.000	2048	9.995	0.100	20474	0.000	99.561	9.961	203940.0
90	0.000	5313	9.998	0.100	53120	0.000	99.812	9.983	530302.3
100	0.000	13781	9.999	0.100	137796	0.000	99.920	9.993	1376961.2
180	0.000	28228209	10.000	0.100	********	0.000	100.000	10.000	*********
240	0.000	*********	10.000	0.100	********	0.000	100.000	10.000	*********
360	0.000	*********	10.000	0.100	********	0.000	100.000	10.000	*********

I = 0.12

n	P/F	F/P	P/A	A/P	F/A	A/F	P/G	0.000	F/G
1	0.893	1.120	0.893	1.120	1.000	1.000	0.000	0.000	0.0
2	0.797	1.254	1.690	0.592	2.120	0.472	0.797	0.472	1.0
3	0.712	1.405	2.402	0.416	3.374	0.296	2.221	0.925	3.1
4	0.636	1.574	3.037	0.329	4.779	0.209	4.127	1.359	6.5
5	0.567	1.762	3.605	0.277	6.353	0.157	6.397	1.775	11.3
6	0.507	1.974	4.111	0.243	8.115	0.123	8.930	2.172	17.6
7	0.452	2.211	4.564	0.219	10.089	0.099	11.644	2.551	25.7
8	0.404	2.476	4.968	0.201	12.300	0.081	14.471	2.913	35.8
9	0.361	2.773	5.328	0.188	14.776	0.068	17.356	3.257	48.1
10	0.322	3.106	5.650	0.177	17.549	0.057	20.254	3.585	62.9
11	0.287	3.479	5.938	0.168	20.655	0.048	23.129	3.895	80.5
12	0.257	3.896	6.194	0.161	24.133	0.041	25.952	4.190	101.1
13	0.229	4.363	6.424	0.156	28.029	0.036	28.702	4.468	125.2
14	0.205	4.887	6.628	0.151	32.393	0.031	31.362	4.732	153.3
15	0.183	5.474	6.811	0.147	37.280	0.027	33.920	4.980	185.7
16	0.163	6.130	6.974	0.143	42.753	0.023	36.367	5.215	222.9
17	0.146	6.866	7.120	0.140	48.884	0.020	38.697	5.435	265.7
18	0.130	7.690	7.250	0.138	55.750	0.018	40.908	5.643	314.6
19	0.116	8.613	7.366	0.136	63.440	0.016	42.998	5.838	370.3
20	0.104	9.646	7.469	0.134	72.052	0.014	44.968	6.020	433.8
21	0.093	10.804	7.562	0.132	81.699	0.012	46.819	6.191	505.8
22	0.083	12.100	7.645	0.131	92.503	0.011	48.554	6.351	587.5
23	0.074	13.552	7.718	0.130	104.603	0.010	50.178	6.501	680.0
24	0.066	15.179	7.784	0.128	118.155	0.008	51.693	6.641	784.6
25	0.059	17.000	7.843	0.127	133.334	0.007	53.105	6.771	902.8
26	0.053	19.040	7.896	0.127	150.334	0.007	54.418	6.892	1036.1
27	0.047	21.325	7.943	0.126	169.374	0.006	55.637	7.005	1186.5
28	0.042	23.884	7.984	0.125	190.699	0.005	56.767	7.110	1355.8
29	0.037	26.750	8.022	0.125	214.583	0.005	57.814	7.207	1546.5
30	0.033	29.960	8.055	0.124	241.333	0.004	58.782	7.297	1761.1
31	0.030	33.555	8.085	0.124	271.293	0.004	59.676	7.381	2002.4
32	0.027	37.582	8.112	0.123	304.848	0.003	60.501	7.459	2273.7
33	0.024	42.092	8.135	0.123	342.429	0.003	61.261	7.530	2578.6
34	0.021	47.143	8.157	0.123	384.521	0.003	61.961	7.596	2921.0
35	0.019	52.800	8.176	0.122	431.663	0.002	62.605	7.658	3305.5
40	0.011	93.051	8.244	0.121	767.091	0.001	65.116	7.899	6059.1
45	0.006	163.988	8.283	0.121	1358.230	0.001	66.734	8.057	10943.6
50	0.003	289.002	8.304	0.120	2400.018	0.000	67.762	8.160	19583.5
60	0.001	897.597	8.324	0.120	7471.641	0.000	68.810	8.266	61763.7
70	0.000	2787.800	8.330	0.120	23223	0.000	69.210	8.308	192944.4
80	0.000	8658.483	8.332	0.120	72146	0.000	69.359	8.324	600547.4
90	0.000	26892	8.333	0.120	224091	0.000	69.414	8.330	1866676
100	0.000	83522	8.333	0.120	696011	0.000	69.434	8.332	5799255
180	0.000	*********	8.333	0.120	********	0.000	69.444	8.333	*********
240	0.000	*********	8.333	0.120	********	0.000	69.444	8.333	*********
360	0.000	*********	8.333	0.120	********	0.000	69.444	8.333	*********

I = 0.14

n	P/F	F/P	P/A	A/P	F/A	A/F	P/G	0.000	F/G
1	0.877	1.140	0.877	1.140	1.000	1.000	0.000	0.000	0.0
2	0.769	1.300	1.647	0.607	2.140	0.467	0.769	0.467	1.0
3	0.675	1.482	2.322	0.431	3.440	0.291	2.119	0.913	3.1
4	0.592	1.689	2.914	0.343	4.921	0.203	3.896	1.337	6.6
5	0.519	1.925	3.433	0.291	6.610	0.151	5.973	1.740	11.5
6	0.456	2.195	3.889	0.257	8.536	0.117	8.251	2.122	18.1
7	0.400	2.502	4.288	0.233	10.730	0.093	10.649	2.483	26.6
8	0.351	2.853	4.639	0.216	13.233	0.076	13.103	2.825	37.4
9	0.308	3.252	4.946	0.202	16.085	0.062	15.563	3.146	50.6
10	0.270	3.707	5.216	0.192	19.337	0.052	17.991	3.449	66.7
11	0.237	4.226	5.453	0.183	23.045	0.043	20.357	3.733	86.0
12	0.208	4.818	5.660	0.177	27.271	0.037	22.640	4.000	109.1
13	0.182	5.492	5.842	0.171	32.089	0.031	24.825	4.249	136.3
14	0.160	6.261	6.002	0.167	37.581	0.027	26.901	4.482	168.4
15	0.140	7.138	6.142	0.163	43.842	0.023	28.862	4.699	206.0
16	0.123	8.137	6.265	0.160	50.980	0.020	30.706	4.901	249.9
17	0.108	9.276	6.373	0.157	59.118	0.017	32.430	5.089	300.8
18	0.095	10.575	6.467	0.155	68.394	0.015	34.038	5.263	360.0
19	0.083	12.056	6.550	0.153	78.969	0.013	35.531	5.424	428.4
20	0.073	13.743	6.623	0.151	91.025	0.011	36.914	5.573	507.3
21	0.064	15.668	6.687	0.150	104.768	0.010	38.190	5.711	598.3
22	0.056	17.861	6.743	0.148	120.436	0.008	39.366	5.838	703.1
23	0.049	20.362	6.792	0.147	138.297	0.007	40.446	5.955	823.6
24	0.043	23.212	6.835	0.146	158.659	0.006	41.437	6.062	961.8
25	0.038	26.462	6.873	0.145	181.871	0.005	42.344	6.161	1120.5
26	0.033	30.167	6.906	0.145	208.333	0.005	43.173	6.251	1302.4
27	0.029	34.390	6.935	0.144	238.499	0.004	43.929	6.334	1510.7
28	0.026	39.204	6.961	0.144	272.889	0.004	44.618	6.410	1749.2
29	0.022	44.693	6.983	0.143	312.094	0.003	45.244	6.479	2022.1
30	0.020	50.950	7.003	0.143	356.787	0.003	45.813	6.542	2334.2
31	0.017	58.083	7.020	0.142	407.737	0.002	46.330	6.600	2691.0
32	0.015	66.215	7.035	0.142	465.820	0.002	46.798	6.652	3098.7
33	0.013	75.485	7.048	0.142	532.035	0.002	47.222	6.700	3564.5
34	0.012	86.053	7.060	0.142	607.520	0.002	47.605	6.743	4096.6
35	0.010	98.100	7.070	0.141	693.573	0.001	47.952	6.782	4704.1
40	0.005	188.884	7.105	0.141	1342.025	0.001	49.238	6.930	9300.2
45	0.003	363.679	7.123	0.140	2590.565	0.000	49.996	7.019	18182.6
50	0.001	700.233	7.133	0.140	4994.521	0.000	50.438	7.071	35318.0
60	0.000	2595.919	7.140	0.140	18535	0.000	50.836	7.120	131965.2
70	0.000	9623.645	7.142	0.140	68733	0.000	50.963	7.136	490451.3
80	0.000	35677	7.143	0.140	254828	0.000	51.003	7.141	1819632
90	0.000	132262	7.143	0.140	944725	0.000	51.015	7.142	6747391
100	0.000	490326	7.143	0.140	3502323	0.000	51.019	7.143	25015879
180	0.000	********	7.143	0.140	********	0.000	51.020	7.143	********
240	0.000	********	7.143	0.140	********	0.000	51.020	7.143	********
360	0.000	********	7.143	0.140	********	0.000	51.020	7.143	********

I = 0.15

n	P/F	F/P	P/A	A/P	F/A	A/F	P/G	0.000	F/G
1	0.870	1.150	0.870	1.150	1.000	1.000	-0.000	-0.000	-0.0
2	0.756	1.323	1.626	0.615	2.150	0.465	0.756	0.465	1.0
3	0.658	1.521	2.283	0.438	3.472	0.288	2.071	0.907	3.1
4	0.572	1.749	2.855	0.350	4.993	0.200	3.786	1.326	6.6
5	0.497	2.011	3.352	0.298	6.742	0.148	5.775	1.723	11.6
6	0.432	2.313	3.784	0.264	8.754	0.114	7.937	2.097	18.4
7	0.376	2.660	4.160	0.240	11.067	0.090	10.192	2.450	27.1
8	0.327	3.059	4.487	0.223	13.727	0.073	12.481	2.781	38.2
9	0.284	3.518	4.772	0.210	16.786	0.060	14.755	3.092	51.9
10	0.247	4.046	5.019	0.199	20.304	0.049	16.979	3.383	68.7
11	0.215	4.652	5.234	0.191	24.349	0.041	19.129	3.655	89.0
12	0.187	5.350	5.421	0.184	29.002	0.034	21.185	3.908	113.3
13	0.163	6.153	5.583	0.179	34.352	0.029	23.135	4.144	142.3
14	0.141	7.076	5.724	0.175	40.505	0.025	24.972	4.362	176.7
15	0.123	8.137	5.847	0.171	47.580	0.021	26.693	4.565	217.2
16	0.107	9.358	5.954	0.168	55.717	0.018	28.296	4.752	264.8
17	0.093	10.761	6.047	0.165	65.075	0.015	29.783	4.925	320.5
18	0.081	12.375	6.128	0.163	75.836	0.013	31.156	5.084	385.6
19	0.070	14.232	6.198	0.161	88.212	0.011	32.421	5.231	461.4
20	0.061	16.367	6.259	0.160	102.444	0.010	33.582	5.365	549.6
21	0.053	18.822	6.312	0.158	118.810	0.008	34.645	5.488	652.1
22	0.046	21.645	6.359	0.157	137.632	0.007	35.615	5.601	770.9
23	0.040	24.891	6.399	0.156	159.276	0.006	36.499	5.704	908.5
24	0.035	28.625	6.434	0.155	184.168	0.005	37.302	5.798	1067.8
25	0.030	32.919	6.464	0.155	212.793	0.005	38.031	5.883	1252.0
26	0.026	37.857	6.491	0.154	245.712	0.004	38.692	5.961	1464.7
27	0.023	43.535	6.514	0.154	283.569	0.004	39.289	6.032	1710.5
28	0.020	50.066	6.534	0.153	327.104	0.003	39.828	6.096	1994.0
29	0.017	57.575	6.551	0.153	377.170	0.003	40.315	6.154	2321.1
30	0.015	66.212	6.566	0.152	434.745	0.002	40.753	6.207	2698.3
31	0.013	76.144	6.579	0.152	500.957	0.002	41.147	6.254	3133.0
32	0.011	87.565	6.591	0.152	577.100	0.002	41.501	6.297	3634.0
33	0.010	100.700	6.600	0.152	664.666	0.002	41.818	6.336	4211.1
34	0.009	115.805	6.609	0.151	765.365	0.001	42.103	6.371	4875.8
35	0.008	133.176	6.617	0.151	881.170	0.001	42.359	6.402	5641.1
40	0.004	267.864	6.642	0.151	1779.090	0.001	43.283	6.517	11593.9
45	0.002	538.769	6.654	0.150	3585.128	0.000	43.805	6.583	23600.9
50	0.001	1083.657	6.661	0.150	7217.716	0.000	44.096	6.620	47784.8
60	0.000	4383.999	6.665	0.150	29220	0.000	44.343	6.653	194399.9
70	0.000	17736	6.666	0.150	118231	0.000	44.416	6.663	787743.1
80	0.000	71751	6.667	0.150	478333	0.000	44.436	6.666	3188350
90	0.000	290272	6.667	0.150	1935142	0.000	44.442	6.666	12900348
100	0.000	1174313	6.667	0.150	7828750	0.000	44.444	6.667	52190998
180	0.000	*********	6.667	0.150	********	0.000	44.444	6.667	*********
240	0.000	*********	6.667	0.150	********	0.000	44.444	6.667	*********
360	0.000	*********	6.667	0.150	********	0.000	44.444	6.667	*********

I=0.16

n	P/F	F/P	P/A	A/P	F/A	A/F	P/G	0.000	F/G
1	0.862	1.160	0.862	1.160	1.000	1.000	-0.000	-0.000	-0.0
2	0.743	1.346	1.605	0.623	2.160	0.463	0.743	0.463	1.0
3	0.641	1.561	2.246	0.445	3.506	0.285	2.024	0.901	3.2
4	0.552	1.811	2.798	0.357	5.066	0.197	3.681	1.316	6.7
5	0.476	2.100	3.274	0.305	6.877	0.145	5.586	1.706	11.7
6	0.410	2.436	3.685	0.271	8.977	0.111	7.638	2.073	18.6
7	0.354	2.826	4.039	0.248	11.414	0.088	9.761	2.417	27.6
8	0.305	3.278	4.344	0.230	14.240	0.070	11.896	2.739	39.0
9	0.263	3.803	4.607	0.217	17.519	0.057	14.000	3.039	53.2
10	0.227	4.411	4.833	0.207	21.321	0.047	16.040	3.319	70.8
11	0.195	5.117	5.029	0.199	25.733	0.039	17.994	3.578	92.1
12	0.168	5.936	5.197	0.192	30.850	0.032	19.847	3.819	117.8
13	0.145	6.886	5.342	0.187	36.786	0.027	21.590	4.041	148.7
14	0.125	7.988	5.468	0.183	43.672	0.023	23.217	4.246	185.4
15	0.108	9.266	5.575	0.179	51.660	0.019	24.728	4.435	229.1
16	0.093	10.748	5.668	0.176	60.925	0.016	26.124	4.609	280.8
17	0.080	12.468	5.749	0.174	71.673	0.014	27.407	4.768	341.7
18	0.069	14.463	5.818	0.172	84.141	0.012	28.583	4.913	413.4
19	0.060	16.777	5.877	0.170	98.603	0.010	29.656	5.046	497.5
20	0.051	19.461	5.929	0.169	115.380	0.009	30.632	5.167	596.1
21	0.044	22.574	5.973	0.167	134.841	0.007	31.518	5.277	711.5
22	0.038	26.186	6.011	0.166	157.415	0.006	32.320	5.377	846.3
23	0.033	30.376	6.044	0.165	183.601	0.005	33.044	5.467	1003.8
24	0.028	35.236	6.073	0.165	213.978	0.005	33.697	5.549	1187.4
25	0.024	40.874	6.097	0.164	249.214	0.004	34.284	5.623	1401.3
26	0.021	47.414	6.118	0.163	290.088	0.003	34.811	5.690	1650.6
27	0.018	55.000	6.136	0.163	337.502	0.003	35.284	5.750	1940.6
28	0.016	63.800	6.152	0.163	392.503	0.003	35.707	5.804	2278.1
29	0.014	74.009	6.166	0.162	456.303	0.002	36.086	5.853	2670.6
30	0.012	85.850	6.177	0.162	530.312	0.002	36.423	5.896	3126.9
31	0.010	99.586	6.187	0.162	616.162	0.002	36.725	5.936	3657.3
32	0.009	115.520	6.196	0.161	715.747	0.001	36.993	5.971	4273.4
33	0.007	134.003	6.203	0.161	831.267	0.001	37.232	6.002	4989.2
34	0.006	155.443	6.210	0.161	965.270	0.001	37.444	6.030	5820.4
35	0.006	180.314	6.215	0.161	1120.713	0.001	37.633	6.055	6785.7
40	0.003	378.721	6.233	0.160	2360.757	0.000	38.299	6.144	14504.7
45	0.001	795.444	6.242	0.160	4965.274	0.000	38.660	6.193	30751.7
50	0.001	1670.704	6.246	0.160	10436	0.000	38.852	6.220	64910.3
60	0.000	7370.201	6.249	0.160	46058	0.000	39.006	6.242	287484.4
70	0.000	32513	6.250	0.160	203201	0.000	39.048	6.248	1269569
80	0.000	143430	6.250	0.160	896429	0.000	39.059	6.249	5602184
90	0.000	632731	6.250	0.160	3954562	0.000	39.062	6.250	24715448
100	0.000	2791251	6.250	0.160	17445314	0.000	39.062	6.250	109032586
180	0.000	********	6.250	0.160	********	0.000	39.062	6.250	********
180	0.000	********	6.250	0.160	********	0.000	39.062	6.250	********
240	0.000	********	6.250	0.160	********	0.000	39.062	6.250	********

I=0.18

n	P/F	F/P	P/A	A/P	F/A	A/F	P/G	0.000	F/G
1	0.847	1.180	0.847	1.180	1.000	1.000	0.000	0.000	0.0
2	0.718	1.392	1.566	0.639	2.180	0.459	0.718	0.459	1.0
3	0.609	1.643	2.174	0.460	3.572	0.280	1.935	0.890	3.2
4	0.516	1.939	2.690	0.372	5.215	0.192	3.483	1.295	6.8
5	0.437	2.288	3.127	0.320	7.154	0.140	5.231	1.673	12.0
6	0.370	2.700	3.498	0.286	9.442	0.106	7.083	2.025	19.1
7	0.314	3.185	3.812	0.262	12.142	0.082	8.967	2.353	28.6
8	0.266	3.759	4.078	0.245	15.327	0.065	10.829	2.656	40.7
9	0.225	4.435	4.303	0.232	19.086	0.052	12.633	2.936	56.0
10	0.191	5.234	4.494	0.223	23.521	0.043	14.352	3.194	75.1
11	0.162	6.176	4.656	0.215	28.755	0.035	15.972	3.430	98.6
12	0.137	7.288	4.793	0.209	34.931	0.029	17.481	3.647	127.4
13	0.116	8.599	4.910	0.204	42.219	0.024	18.877	3.845	162.3
14	0.099	10.147	5.008	0.200	50.818	0.020	20.158	4.025	204.5
15	0.084	11.974	5.092	0.196	60.965	0.016	21.327	4.189	255.4
16	0.071	14.129	5.162	0.194	72.939	0.014	22.389	4.337	316.3
17	0.060	16.672	5.222	0.191	87.068	0.011	23.348	4.471	389.3
18	0.051	19.673	5.273	0.190	103.740	0.010	24.212	4.592	476.3
19	0.043	23.214	5.316	0.188	123.414	0.008	24.988	4.700	580.1
20	0.037	27.393	5.353	0.187	146.628	0.007	25.681	4.798	703.5
21	0.031	32.324	5.384	0.186	174.021	0.006	26.300	4.885	850.1
22	0.026	38.142	5.410	0.185	206.345	0.005	26.851	4.963	1024.1
23	0.022	45.008	5.432	0.184	244.487	0.004	27.339	5.033	1230.5
24	0.019	53.109	5.451	0.183	289.494	0.003	27.772	5.095	1475.0
25	0.016	62.669	5.467	0.183	342.603	0.003	28.155	5.150	1764.5
26	0.014	73.949	5.480	0.182	405.272	0.002	28.494	5.199	2107.1
27	0.011	87.260	5.492	0.182	479.221	0.002	28.791	5.243	2512.3
28	0.010	102.967	5.502	0.182	566.481	0.002	29.054	5.281	2991.6
29	0.008	121.501	5.510	0.181	669.447	0.001	29.284	5.315	3558.0
30	0.007	143.371	5.517	0.181	790.948	0.001	29.486	5.345	4227.5
31	0.006	169.177	5.523	0.181	934.319	0.001	29.664	5.371	5018.4
32	0.005	199.629	5.528	0.181	1103.496	0.001	29.819	5.394	5952.8
33	0.004	235.563	5.532	0.181	1303.125	0.001	29.955	5.415	7056.3
34	0.004	277.964	5.536	0.181	1538.688	0.001	30.074	5.433	8359.4
35	0.003	327.997	5.539	0.181	1816.652	0.001	30.177	5.449	9898.1
40	0.001	750.378	5.548	0.180	4163.213	0.000	30.527	5.502	22906.7
45	0.001	1716.684	5.552	0.180	9531.577	0.000	30.701	5.529	52703.2
50	0.000	3927.357	5.554	0.180	21813	0.000	30.786	5.543	120906.1
60	0.000	20555	5.555	0.180	114190	0.000	30.846	5.553	634053.7
70	0.000	107582	5.556	0.180	597673	0.000	30.860	5.555	3320019
80	0.000	563068	5.556	0.180	3128148	0.000	30.863	5.555	17378156
90	0.000	2947004	5.556	0.180	16372236	0.000	30.864	5.556	90956369
100	0.000	15424132	5.556	0.180	85689616	0.000	30.864	5.556	476052867
180	0.000	*********	5.556	0.180	********	0.000	30.864	5.556	*********
240	0.000	*********	5.556	0.180	********	0.000	30.864	5.556	*********
360	0.000	*********	5.556	0.180	********	0.000	30.864	5.556	*********

I = 0.2

n	P/F	F/P	P/A	A/P	F/A	A/F	P/G	0.000	F/G
1	0.833	1.200	0.833	1.200	1.000	1.000	0.000	0.000	0.0
2	0.694	1.440	1.528	0.655	2.200	0.455	0.694	0.455	1.0
3	0.579	1.728	2.106	0.475	3.640	0.275	1.852	0.879	3.2
4	0.482	2.074	2.589	0.386	5.368	0.186	3.299	1.274	6.8
5	0.402	2.488	2.991	0.334	7.442	0.134	4.906	1.641	12.2
6	0.335	2.986	3.326	0.301	9.930	0.101	6.581	1.979	19.6
7	0.279	3.583	3.605	0.277	12.916	0.077	8.255	2.290	29.6
8	0.233	4.300	3.837	0.261	16.499	0.061	9.883	2.576	42.5
9	0.194	5.160	4.031	0.248	20.799	0.048	11.434	2.836	59.0
10	0.162	6.192	4.192	0.239	25.959	0.039	12.887	3.074	79.8
11	0.135	7.430	4.327	0.231	32.150	0.031	14.233	3.289	105.8
12	0.112	8.916	4.439	0.225	39.581	0.025	15.467	3.484	137.9
13	0.093	10.699	4.533	0.221	48.497	0.021	16.588	3.660	177.5
14	0.078	12.839	4.611	0.217	59.196	0.017	17.601	3.817	226.0
15	0.065	15.407	4.675	0.214	72.035	0.014	18.509	3.959	285.2
16	0.054	18.488	4.730	0.211	87.442	0.011	19.321	4.085	357.2
17	0.045	22.186	4.775	0.209	105.931	0.009	20.042	4.198	444.7
18	0.038	26.623	4.812	0.208	128.117	0.008	20.680	4.298	550.6
19	0.031	31.948	4.843	0.206	154.740	0.006	21.244	4.386	678.7
20	0.026	38.338	4.870	0.205	186.688	0.005	21.739	4.464	833.4
21	0.022	46.005	4.891	0.204	225.026	0.004	22.174	4.533	1020.1
22	0.018	55.206	4.909	0.204	271.031	0.004	22.555	4.594	1245.2
23	0.015	66.247	4.925	0.203	326.237	0.003	22.887	4.647	1516.2
24	0.013	79.497	4.937	0.203	392.484	0.003	23.176	4.694	1842.4
25	0.010	95.396	4.948	0.202	471.981	0.002	23.428	4.735	2234.9
26	0.009	114.475	4.956	0.202	567.377	0.002	23.646	4.771	2706.9
27	0.007	137.371	4.964	0.201	681.853	0.001	23.835	4.802	3274.3
28	0.006	164.845	4.970	0.201	819.223	0.001	23.999	4.829	3956.1
29	0.005	197.814	4.975	0.201	984.068	0.001	24.141	4.853	4775.3
30	0.004	237.376	4.979	0.201	1181.882	0.001	24.263	4.873	5759.4
31	0.004	284.852	4.982	0.201	1419.258	0.001	24.368	4.891	6941.3
32	0.003	341.822	4.985	0.201	1704.109	0.001	24.459	4.906	8360.5
33	0.002	410.186	4.988	0.200	2045.931	0.000	24.537	4.919	10064.7
34	0.002	492.224	4.990	0.200	2456.118	0.000	24.604	4.931	12110.6
35	0.002	590.668	4.992	0.200	2948.341	0.000	24.661	4.941	14566.7
40	0.001	1469.772	4.997	0.200	7343.858	0.000	24.847	4.973	36519.3
45	0.000	3657.262	4.999	0.200	18281	0.000	24.932	4.988	91181.5
50	0.000	9100.438	4.999	0.200	45497	0.000	24.970	4.995	227236.0
60	0.000	56348	5.000	0.200	281733	0.000	24.994	4.999	1408363
70	0.000	348889	5.000	0.200	1744440	0.000	24.999	5.000	8721849
80	0.000	2160228	5.000	0.200	10801137	0.000	25.000	5.000	54005287
90	0.000	13375565	5.000	0.200	66877821	0.000	25.000	5.000	334388656
100	0.000	82817975	5.000	0.200	********	0.000	25.000	5.000	*********
180	0.000	*********	5.000	0.200	********	0.000	25.000	5.000	*********
240	0.000	*********	5.000	0.200	********	0.000	25.000	5.000	*********
360	0.000	*********	5.000	0.200	********	0.000	25.000	5.000	*********

n	P/F	F/P	P/A	A/P	F/A	A/F	P/G	0.000	F/G
1	0.800	1.250	0.800	1.250	1.000	1.000	0.000	0.000	0.0
2	0.640	1.563	1.440	0.694	2.250	0.444	0.640	0.444	1.0
3	0.512	1.953	1.952	0.512	3.813	0.262	1.664	0.852	3.2
4	0.410	2.441	2.362	0.423	5.766	0.173	2.893	1.225	7.1
5	0.328	3.052	2.689	0.372	8.207	0.122	4.204	1.563	12.8
6	0.262	3.815	2.951	0.339	11.259	0.089	5.514	1.868	21.0
7	0.210	4.768	3.161	0.316	15.073	0.066	6.773	2.142	32.3
8	0.168	5.960	3.329	0.300	19.842	0.050	7.947	2.387	47.4
9	0.134	7.451	3.463	0.289	25.802	0.039	9.021	2.605	67.2
10	0.107	9.313	3.571	0.280	33.253	0.030	9.987	2.797	93.0
11	0.086	11.642	3.656	0.273	42.566	0.023	10.846	2.966	126.3
12	0.069	14.552	3.725	0.268	54.208	0.018	11.602	3.115	168.8
13	0.055	18.190	3.780	0.265	68.760	0.015	12.262	3.244	223.0
14	0.044	22.737	3.824	0.262	86.949	0.012	12.833	3.356	291.8
15	0.035	28.422	3.859	0.259	109.687	0.009	13.326	3.453	378.7
16	0.028	35.527	3.887	0.257	138.109	0.007	13.748	3.537	488.4
17	0.023	44.409	3.910	0.256	173.636	0.006	14.108	3.608	626.5
18	0.018	55.511	3.928	0.255	218.045	0.005	14.415	3.670	800.2
19	0.014	69.389	3.942	0.254	273.556	0.004	14.674	3.722	1018.2
20	0.012	86.736	3.954	0.253	342.945	0.003	14.893	3.767	1291.8
21	0.009	108.420	3.963	0.252	429.681	0.002	15.078	3.805	1634.7
22	0.007	135.525	3.970	0.252	538.101	0.002	15.233	3.836	2064.4
23	0.006	169.407	3.976	0.251	673.626	0.001	15.362	3.863	2602.5
24	0.005	211.758	3.981	0.251	843.033	0.001	15.471	3.886	3276.1
25	0.004	264.698	3.985	0.251	1054.791	0.001	15.562	3.905	4119.2
26	0.003	330.872	3.988	0.251	1319.489	0.001	15.637	3.921	5174.0
27	0.002	413.590	3.990	0.251	1650.361	0.001	15.700	3.935	6493.4
28	0.002	516.988	3.992	0.250	2063.952	0.000	15.752	3.946	8143.8
29	0.002	646.235	3.994	0.250	2580.939	0.000	15.796	3.955	10207.8
30	0.001	807.794	3.995	0.250	3227.174	0.000	15.832	3.963	12788.7
31	0.001	1009.742	3.996	0.250	4034.968	0.000	15.861	3.969	16015.9
32	0.001	1262.177	3.997	0.250	5044.710	0.000	15.886	3.975	20050.8
33	0.001	1577.722	3.997	0.250	6306.887	0.000	15.906	3.979	25095.5
34	0.001	1972.152	3.998	0.250	7884.609	0.000	15.923	3.983	31402.4
35	0.000	2465.190	3.998	0.250	9856.761	0.000	15.937	3.986	39287.0
40	0.000	7523.164	3.999	0.250	30088.66	0.000	15.977	3.995	120194.6
45	0.000	22958.874	4.000	0.250	91831.50	0.000	15.991	3.998	367146.0
50	0.000	70064.923	4.000	0.250	280255.7	0.000	15.997	3.999	1120822.8
60	0.000	652530.45	4.000	0.250	2610118	0.000	16.000	4.000	10440231
70	0.000	6077163	4.000	0.250	24308649	0.000	16.000	4.000	97234318
80	0.000	56597994	4.000	0.250	********	0.000	16.000	4.000	905567572
90	0.000	********	4.000	0.250	********	0.000	16.000	4.000	********
100	0.000	********	4.000	0.250	********	0.000	16.000	4.000	********
180	0.000	********	4.000	0.250	********	0.000	16.000	4.000	********
240	0.000	********	4.000	0.250	********	0.000	16.000	4.000	********
360	0.000	********	4.000	0.250	********	0.000	16.000	4.000	********

I = 0.3

n	P/F	F/P	P/A	A/P	F/A	A/F	P/G	0.000	F/G
1	0.769	1.300	0.769	1.300	1.000	1.000	0.000	0.000	0.0
2	0.592	1.690	1.361	0.735	2.300	0.435	0.592	0.435	1.0
3	0.455	2.197	1.816	0.551	3.990	0.251	1.502	0.827	3.3
4	0.350	2.856	2.166	0.462	6.187	0.162	2.552	1.178	7.3
5	0.269	3.713	2.436	0.411	9.043	0.111	3.630	1.490	13.5
6	0.207	4.827	2.643	0.378	12.756	0.078	4.666	1.765	22.5
7	0.159	6.275	2.802	0.357	17.583	0.057	5.622	2.006	35.3
8	0.123	8.157	2.925	0.342	23.858	0.042	6.480	2.216	52.9
9	0.094	10.604	3.019	0.331	32.015	0.031	7.234	2.396	76.7
10	0.073	13.786	3.092	0.323	42.619	0.023	7.887	2.551	108.7
11	0.056	17.922	3.147	0.318	56.405	0.018	8.445	2.683	151.4
12	0.043	23.298	3.190	0.313	74.327	0.013	8.917	2.795	207.8
13	0.033	30.288	3.223	0.310	97.625	0.010	9.314	2.889	282.1
14	0.025	39.374	3.249	0.308	127.913	0.008	9.644	2.969	379.7
15	0.020	51.186	3.268	0.306	167.286	0.006	9.917	3.034	507.6
16	0.015	66.542	3.283	0.305	218.472	0.005	10.143	3.089	674.9
17	0.012	86.504	3.295	0.304	285.014	0.004	10.328	3.135	893.4
18	0.009	112.455	3.304	0.303	371.518	0.003	10.479	3.172	1178.4
19	0.007	146.192	3.311	0.302	483.973	0.002	10.602	3.202	1549.9
20	0.005	190.050	3.316	0.302	630.165	0.002	10.702	3.228	2033.9
21	0.004	247.065	3.320	0.301	820.215	0.001	10.783	3.248	2664.1
22	0.003	321.184	3.323	0.301	1067.280	0.001	10.848	3.265	3484.3
23	0.002	417.539	3.325	0.301	1388.464	0.001	10.901	3.278	4551.5
24	0.002	542.801	3.327	0.301	1806.003	0.001	10.943	3.289	5940.0
25	0.001	705.641	3.329	0.300	2348.803	0.000	10.977	3.298	7746.0
26	0.001	917.333	3.330	0.300	3054.444	0.000	11.005	3.305	10094.8
27	0.001	1192.533	3.331	0.300	3971.778	0.000	11.026	3.311	13149.3
28	0.001	1550.293	3.331	0.300	5164.311	0.000	11.044	3.315	17121.0
29	0.000	2015.381	3.332	0.300	6714.604	0.000	11.058	3.319	22285.3
30	0.000	2619.996	3.332	0.300	8729.985	0.000	11.069	3.322	29000.0
31	0.000	3405.994	3.332	0.300	11350	0.000	11.078	3.324	37729.9
32	0.000	4427.793	3.333	0.300	14756	0.000	11.085	3.326	49079.9
33	0.000	5756.130	3.333	0.300	19184	0.000	11.090	3.328	63835.9
34	0.000	7482.970	3.333	0.300	24940	0.000	11.094	3.329	83019.7
35	0.000	9727.860	3.333	0.300	32423	0.000	11.098	3.330	107959.6
40	0.000	36118.865	3.333	0.300	120393	0.000	11.107	3.332	401176.3
45	0.000	134107	3.333	0.300	447019	0.000	11.110	3.333	1489914.6
50	0.000	497929	3.333	0.300	1659761	0.000	11.111	3.333	5532369.1
60	0.000	6864377	3.333	0.300	22881254	0.000	11.111	3.333	76270646
70	0.000	94631268	3.333	0.300	********	0.000	11.111	3.333	*********
80	0.000	********	3.333	0.300	********	0.000	11.111	3.333	*********
90	0.000	********	3.333	0.300	********	0.000	11.111	3.333	*********
100	0.000	********	3.333	0.300	********	0.000	11.111	3.333	*********
180	0.000	********	3.333	0.300	********	0.000	11.111	3.333	*********
240	0.000	********	3.333	0.300	********	0.000	11.111	3.333	*********
360	0.000	********	3.333	0.300	********	0.000	11.111	3.333	*********

I =0.4

n	P/F	F/P	P/A	A/P	F/A	A/F	P/G	A/G	F/G
1	0.714	1.400	0.714	1.400	1.000	1.000	-0.000	-0.000	-0.0
2	0.510	1.960	1.224	0.817	2.400	0.417	0.510	0.417	1.0
3	0.364	2.744	1.589	0.629	4.360	0.229	1.239	0.780	3.4
4	0.260	3.842	1.849	0.541	7.104	0.141	2.020	1.092	7.8
5	0.186	5.378	2.035	0.491	10.946	0.091	2.764	1.358	14.9
6	0.133	7.530	2.168	0.461	16.324	0.061	3.428	1.581	25.8
7	0.095	10.541	2.263	0.442	23.853	0.042	3.997	1.766	42.1
8	0.068	14.758	2.331	0.429	34.395	0.029	4.471	1.919	66.0
9	0.048	20.661	2.379	0.420	49.153	0.020	4.858	2.042	100.4
10	0.035	28.925	2.414	0.414	69.814	0.014	5.170	2.142	149.5
11	0.025	40.496	2.438	0.410	98.739	0.010	5.417	2.221	219.3
12	0.018	56.694	2.456	0.407	139.235	0.007	5.611	2.285	318.1
13	0.013	79.371	2.469	0.405	195.929	0.005	5.762	2.334	457.3
14	0.009	111.120	2.478	0.404	275.300	0.004	5.879	2.373	653.3
15	0.006	155.568	2.484	0.403	386.420	0.003	5.969	2.403	928.6
16	0.005	217.795	2.489	0.402	541.988	0.002	6.038	2.426	1315.0
17	0.003	304.913	2.492	0.401	759.784	0.001	6.090	2.444	1857.0
18	0.002	426.879	2.494	0.401	1064.697	0.001	6.130	2.458	2616.7
19	0.002	597.630	2.496	0.401	1491.576	0.001	6.160	2.468	3681.4
20	0.001	836.683	2.497	0.400	2089.206	0.000	6.183	2.476	5173.0
21	0.001	1171.356	2.498	0.400	2925.889	0.000	6.200	2.482	7262.2
22	0.001	1639.898	2.498	0.400	4097.245	0.000	6.213	2.487	10188.1
23	0.000	2295.857	2.499	0.400	5737.142	0.000	6.222	2.490	14285.4
24	0.000	3214.200	2.499	0.400	8032.999	0.000	6.229	2.493	20022.5
25	0.000	4499.880	2.499	0.400	11247	0.000	6.235	2.494	28055.5
26	0.000	6299.831	2.500	0.400	15747	0.000	6.239	2.496	39302.7
27	0.000	8819.764	2.500	0.400	22047	0.000	6.242	2.497	55049.8
28	0.000	12347.670	2.500	0.400	30867	0.000	6.244	2.498	77096.7
29	0.000	17286.737	2.500	0.400	43214	0.000	6.245	2.498	107963.4
30	0.000	24201.432	2.500	0.400	60501	0.000	6.247	2.499	151177.7
31	0.000	33882.005	2.500	0.400	84703	0.000	6.248	2.499	211678.8
32	0.000	47434.807	2.500	0.400	118585	0.000	6.248	2.499	296381.3
33	0.000	66408.730	2.500	0.400	166019	0.000	6.249	2.500	414965.8
34	0.000	92972.223	2.500	0.400	232428	0.000	6.249	2.500	580985.1
35	0.000	130161	2.500	0.400	325400	0.000	6.249	2.500	813413.2
40	0.000	700038	2.500	0.400	1750092	0.000	6.250	2.500	4375129.4
45	0.000	3764971	2.500	0.400	9412424	0.000	6.250	2.500	23530948
50	0.000	20248916	2.500	0.400	50622288	0.000	6.250	2.500	126555595
60	0.000	********	2.500	0.400	********	0.000	6.250	2.500	********
70	0.000	********	2.500	0.400	********	0.000	6.250	2.500	********
80	0.000	********	2.500	0.400	********	0.000	6.250	2.500	********
90	0.000	********	2.500	0.400	********	0.000	6.250	2.500	********
100	0.000	********	2.500	0.400	********	0.000	6.250	2.500	********
180	0.000	********	2.500	0.400	********	0.000	6.250	2.500	********
240	0.000	********	2.500	0.400	********	0.000	6.250	2.500	********
360	0.000	********	2.500	0.400	********	0.000	6.250	2.500	********

216

I = 0.5

n	P/F	F/P	P/A	A/P	F/A	A/F	P/G	A/G	F/G
1	0.667	1.500	0.667	1.500	1.000	1.000	0.000	0.000	0.0
2	0.444	2.250	1.111	0.900	2.500	0.400	0.444	0.400	1.0
3	0.296	3.375	1.407	0.711	4.750	0.211	1.037	0.737	3.5
4	0.198	5.063	1.605	0.623	8.125	0.123	1.630	1.015	8.3
5	0.132	7.594	1.737	0.576	13.188	0.076	2.156	1.242	16.4
6	0.088	11.391	1.824	0.548	20.781	0.048	2.595	1.423	29.6
7	0.059	17.086	1.883	0.531	32.172	0.031	2.947	1.565	50.3
8	0.039	25.629	1.922	0.520	49.258	0.020	3.220	1.675	82.5
9	0.026	38.443	1.948	0.513	74.887	0.013	3.428	1.760	131.8
10	0.017	57.665	1.965	0.509	113.330	0.009	3.584	1.824	206.7
11	0.012	86.498	1.977	0.506	170.995	0.006	3.699	1.871	320.0
12	0.008	129.746	1.985	0.504	257.493	0.004	3.784	1.907	491.0
13	0.005	194.620	1.990	0.503	387.239	0.003	3.846	1.933	748.5
14	0.003	291.929	1.993	0.502	581.859	0.002	3.890	1.952	1135.7
15	0.002	437.894	1.995	0.501	873.788	0.001	3.922	1.966	1717.6
16	0.002	656.841	1.997	0.501	1311.682	0.001	3.945	1.976	2591.4
17	0.001	985.261	1.998	0.501	1968.523	0.001	3.961	1.983	3903.0
18	0.001	1477.892	1.999	0.500	2953.784	0.000	3.973	1.988	5871.6
19	0.000	2216.838	1.999	0.500	4431.676	0.000	3.981	1.991	8825.4
20	0.000	3325.257	1.999	0.500	6648.513	0.000	3.987	1.994	13257.0
21	0.000	4987.885	2.000	0.500	9973.770	0.000	3.991	1.996	19905.5
22	0.000	7481.828	2.000	0.500	14962	0.000	3.994	1.997	29879.3
23	0.000	11222.741	2.000	0.500	22443	0.000	3.996	1.998	44841.0
24	0.000	16834.112	2.000	0.500	33666	0.000	3.997	1.999	67284.4
25	0.000	25251.168	2.000	0.500	50500	0.000	3.998	1.999	100950.7
26	0.000	37876.752	2.000	0.500	75752	0.000	3.999	1.999	151451.0
27	0.000	56815.129	2.000	0.500	113628	0.000	3.999	2.000	227202.5
28	0.000	85222.693	2.000	0.500	170443	0.000	3.999	2.000	340830.8
29	0.000	127834	2.000	0.500	255666	0.000	4.000	2.000	511274.2
30	0.000	191751	2.000	0.500	383500	0.000	4.000	2.000	766940.2
31	0.000	287627	2.000	0.500	575251	0.000	4.000	2.000	1150440.4
32	0.000	431440	2.000	0.500	862878	0.000	4.000	2.000	1725691.5
33	0.000	647160	2.000	0.500	1294318	0.000	4.000	2.000	2588569.3
34	0.000	970740	2.000	0.500	1941477	0.000	4.000	2.000	3882886.9
35	0.000	1456110	2.000	0.500	2912217	0.000	4.000	2.000	5824364.4
40	0.000	11057332	2.000	0.500	22114663	0.000	4.000	2.000	44229245
45	0.000	83966617	2.000	0.500	********	0.000	4.000	2.000	335866375
50	0.000	********	2.000	0.500	********	0.000	4.000	2.000	*********
60	0.000	********	2.000	0.500	********	0.000	4.000	2.000	*********
70	0.000	********	2.000	0.500	********	0.000	4.000	2.000	*********
80	0.000	********	2.000	0.500	********	0.000	4.000	2.000	*********
90	0.000	********	2.000	0.500	********	0.000	4.000	2.000	*********
100	0.000	********	2.000	0.500	********	0.000	4.000	2.000	*********
180	0.000	********	2.000	0.500	********	0.000	4.000	2.000	*********
240	0.000	********	2.000	0.500	********	0.000	4.000	2.000	*********
360	0.000	********	2.000	0.500	********	0.000	4.000	2.000	*********

Index

About the Author

Dr. Ardalan teaches graduate level courses in
Engineering Economics and Financial Analysis,
Project Management, Technology Management, and
Managerial Economics. He has over thirty years of

experience in academia as well as public and private sector management.
He has worked with many U.S. and international companies at all levels of
management. Dr. Ardalan is the president of AAA Consulting, a
management consulting company in the Washington metropolitan area, and
World Seminars Inc., an organization that conducts management seminars
globally. He has done consulting work and has conducted seminars in the
U.S., Eastern Europe, and the Middle East. Before starting his consulting
company, he was Director of Technology Ventures of the Airspace
Management Systems division at Westinghouse Electronic Systems.

Dr. Ardalan received a Doctor of Science degree in Engineering
Management from The George Washington University, a Master of Science
in Electrical Engineering, and a Bachelor of Science in Mathematics from
the U.S. Navy Post-Graduate School.